T0213302

SpringerBriefs in Environmental Science

SpringerBriefs in Environmental Science present concise summaries of cutting-edge research and practical applications across a wide spectrum of environmental fields, with fast turnaround time to publication. Featuring compact volumes of 50 to 125 pages, the series covers a range of content from professional to academic. Monographs of new material are considered for the SpringerBriefs in Environmental Science series.

Typical topics might include: a timely report of state-of-the-art analytical techniques, a bridge between new research results, as published in journal articles and a contextual literature review, a snapshot of a hot or emerging topic, an in-depth case study or technical example, a presentation of core concepts that students must understand in order to make independent contributions, best practices or protocols to be followed, a series of short case studies/debates highlighting a specific angle.

SpringerBriefs in Environmental Science allow authors to present their ideas and readers to absorb them with minimal time investment. Both solicited and unsolicited manuscripts are considered for publication.

More information about this series at http://www.springer.com/series/8868

Leonel Jorge Ribeiro Nunes ·
Catarina Isabel Rodrigues Meireles ·
Carlos José Pinto Gomes ·
Nuno Manuel Cabral de Almeida Ribeiro

Climate Change Impact on Environmental Variability in the Forest

 Springer

Leonel Jorge Ribeiro Nunes
ICAAM—Instituto de Ciências Agrárias e
Ambientais Mediterrânicas
University of Évora
Évora, Portugal

Catarina Isabel Rodrigues Meireles
ICAAM—Instituto de Ciências Agrárias e
Ambientais Mediterrânicas
University of Évora
Évora, Portugal

Carlos José Pinto Gomes
Departamento de Paisagem
Ambiente e Ordenamento and ICAAM—
Instituto de Ciências Agrárias e Ambientais
Mediterrânicas
University of Évora
Évora, Portugal

Nuno Manuel Cabral de Almeida Ribeiro
Departamento de Fitotecnia
and ICAAM—Instituto de Ciências
Agrárias e Ambientais Mediterrânicas
University of Évora
Évora, Portugal

ISSN 2191-5547   ISSN 2191-5555   (electronic)
SpringerBriefs in Environmental Science
ISBN 978-3-030-34416-0   ISBN 978-3-030-34417-7   (eBook)
https://doi.org/10.1007/978-3-030-34417-7

© The Author(s), under exclusive license to Springer Nature Switzerland AG 2020
This work is subject to copyright. All rights are solely and exclusively licensed by the Publisher, whether the whole or part of the material is concerned, specifically the rights of translation, reprinting, reuse of illustrations, recitation, broadcasting, reproduction on microfilms or in any other physical way, and transmission or information storage and retrieval, electronic adaptation, computer software, or by similar or dissimilar methodology now known or hereafter developed.
The use of general descriptive names, registered names, trademarks, service marks, etc. in this publication does not imply, even in the absence of a specific statement, that such names are exempt from the relevant protective laws and regulations and therefore free for general use.
The publisher, the authors and the editors are safe to assume that the advice and information in this book are believed to be true and accurate at the date of publication. Neither the publisher nor the authors or the editors give a warranty, expressed or implied, with respect to the material contained herein or for any errors or omissions that may have been made. The publisher remains neutral with regard to jurisdictional claims in published maps and institutional affiliations.

This Springer imprint is published by the registered company Springer Nature Switzerland AG
The registered company address is: Gewerbestrasse 11, 6330 Cham, Switzerland

Contents

List of Figures

List of Tables

Chapter 1
Introduction

Abstract Climate change is a hot topic and is discussed almost daily, often being the opening of television news, or the cover of generalist newspapers and magazines. This is an issue that has quickly become known to almost all citizens, thus creating unanimity regarding the existence of climate change. However, there are still negationist currents of thought, which present often unscientific arguments with which to argue. This chapter presents an introduction to the subject of climate change, highlighting the increase in CO_2 concentrations in the atmosphere.

Keywords Climate change · CO_2 concentration · Earth systems · Extreme weather events

According to the Collins English Dictionary, climate can be defined as the set of long-term meteorological conditions prevailing in a given area, conditioned by latitude, relative position of the oceans or continents, altitude, and other conditions (Collins, s.d.).

The study of the climate is a complex field of investigation, in great evolution due mainly to the number of factors that can influence it, such as temperature, precipitation, marine currents, solar radiation, among others. These factors intervene directly in the energy balance of the planet, provoking variations with different temporal scales, from tens to thousands or even millions of years, which make the earth's climate never static. Among the most notable climatic variations that have occurred throughout Earth's history is the approximately 100,000-year cycle of glacial periods, followed by interglacial periods (Füssel 2007).

About 2.6 million years ago, at the beginning of the Pleistocene, large beds of ice that reached several kilometers thick began to appear in the northern hemisphere (Fig. 1.1) (Kurten 2017). These layers of ice advanced during colder glacial periods and receded during warmer interglacial periods. Glacial epochs are powerful evidences of climate change that naturally occurred on Earth in the geological past (Fig. 1.2) (Zagwijn 1989).

Climate change is defined as the global variation of the Earth's climate, due to natural causes, but also to human action (Parmesan and Yohe 2003). Climate change or change occurs at very different times and over all climatic parameters, such as temperature, precipitation, cloudiness, and so on. The designation "greenhouse

© The Author(s), under exclusive license to Springer Nature Switzerland AG 2020

L. J. R. Nunes et al., *Climate Change Impact on Environmental Variability in the Forest*, SpringerBriefs in Environmental Science, https://doi.org/10.1007/978-3-030-34417-7_1

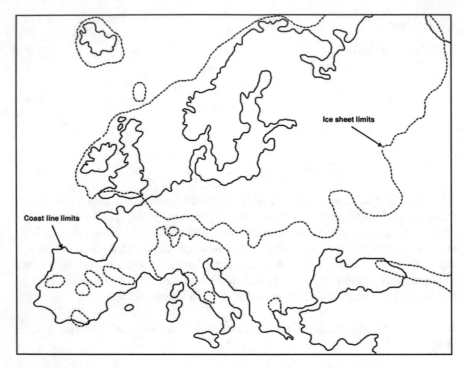

Fig. 1.1 Evolution of the ice layer over Europe during the Pleistocene. Adapted from Varner (2009)

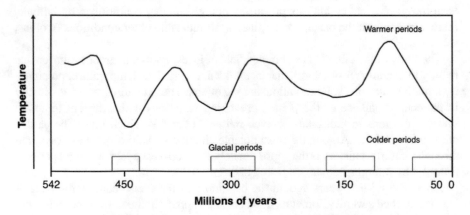

Fig. 1.2 Evolution of the climate in the last 500 million years, with the last three major ice ages indicated. A less severe period of cold occurred during the Jurassic-Cretaceous periods (150 Ma). Adapted from (Wikipedia, s.d.)

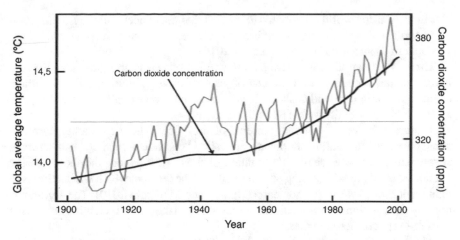

Fig. 1.3 Annual average global temperature. Red bars indicate temperatures above and blue bars indicate temperatures below average temperature for the period 1901–2000. The black line shows the atmospheric concentration of carbon dioxide (CO_2) in parts per million (ppm). Although there is a clear trend of long-term global warming, each year alone does not show a rise in temperature compared to the previous year, and some years show greater changes than others. These year-to-year temperature fluctuations are due to natural processes such as the effects of El Niño, La Niña and the eruption of large volcanoes. Adapted from (NOAA, s.d.)

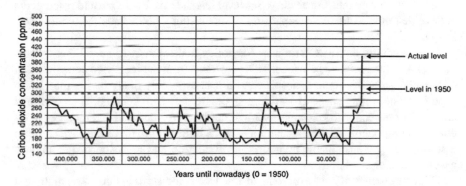

Fig. 1.4 This graph is based on the comparison of atmospheric samples collected in ice cores with more recent direct measurements, and shows how the concentration of carbon dioxide present in the atmosphere has evolved since the beginning of the Industrial Revolution. Adapted from (Petit, s.d.)

effect" applies to the phenomenon of retention of solar radiation by the Earth's atmosphere through a layer of gases called "greenhouse gases" (Fig. 1.3). Without them life as it is known would not be possible, since the planet would be too cold (Moss et al. 2010).

These gases include carbon dioxide, nitrogen oxides and methane, which are released mainly by industry, agriculture and the combustion of fossil fuels (Fig. 1.4).

The industrialized world contributed to the fact that the concentration of these gases increased about 30% during the 20th century, when, without human action, nature was able to balance emissions (Tian et al. 2016).

There is now an almost generalized scientific consensus about the idea that the current model of energy production and consumption is generating a global climate change, which in turn will cause serious impacts both on the planet's environment and on socio-economic systems (Melillo et al. 2014).

There are, however, different currents of this view. These dissenting currents, some based on unscientific arguments, are called "climatic denialism" and "climate skepticism", and they identify the opinion of those who oppose the theory of the phenomenon of global warming or, at least, doubt the fact that which will be the main man responsible for this phenomenon (Dunlap and McCright 2010). The arguments used by the negationists are considered unreliable and the current scientific consensus does not support these theories.

Already in the distant year of 2001, the Third Assessment Report of the Intergovernmental Panel on Climate Change (IPCC) highlighted the evidence provided by observations of physical and biological systems that the regional changes in the climate, more specifically the temperature increases, were affecting the different systems and in different parts of the planet (I. P. O. C. Change 2014, 2015).

The report stated definitively that there was mounting evidence of the existence of climate change and the resulting impacts. However, the temperature increased by about 0.6 °C during the 20th century. Sea level has risen by 10–12 cm and researchers believe that this is due to the expansion of increasingly hot oceans (Mitchum et al. 2017).

More recently, the IPCC summary for policy makers in 2013 stated that "man's influence on the climate is obvious. This is evident from the increased concentration of greenhouse gases in the atmosphere, and it is extremely likely that human influence is the dominant cause of warming since the mid-20th century. The continued emission of greenhouse gas emissions will cause more warming and changes in all climatic components of the planetary system. Limiting climate change will require a substantial and sustainable reduction in production and emission of greenhouse gases" (Stocker 2014).

Climate change affects everyone in the world indiscriminately. The potential impact is huge with predictions of lack of potable water, major changes in food production conditions, and rising mortality rates due to floods, storms, droughts and heat waves (Crate and Nuttall 2016).

Ultimately, climate change is not only an environmental phenomenon, but also with profound economic and social consequences. The poorest countries, which are most ill-equipped to deal with rapid changes, will be those that will suffer the worst consequences (Moss et al. 2010).

Extinction of animals and plants is expected, as habitats will change so sharply that many species will not be able to adapt in time to survive. The World Health Organization (WHO) warned that the health of millions of people could be threatened by increased malaria, malnutrition and waterborne diseases (World Health Organization 2015; Watts et al. 2015).

Portugal, because of its geographical situation and socio-economic characteristics, is very vulnerable to climate change. As a consequence, even if there are uncertainties that do not allow the expected climate change to be quantified with sufficient precision, the information validated so far is sufficient to take immediate action, in accordance with the so-called "Precautionary Principle", which is made reference to Article 3 of the United Nations Framework Convention on Climate Change (UNFCCC) (Henriques 2009).

The inertia, delays and irreversibility of the climate system are very important factors to be taken into account, and the longer it takes to take action, the effects of increasing concentrations of greenhouse gases will be less reversible (Milheiro 2006).

Climate change presents major challenges for the Portuguese forestry sector. The effects of climate change on forest ecosystems in Portugal are already evident in many respects, and the anticipated impacts, according to future climate scenarios, indicate a progressive intensification of these effects as the 21st century progresses, in the distribution of forest formations, structural and functional changes, changes in certain parameters of forest health, greater vulnerability to extreme weather events and fires and modification of the flow of environmental goods and services that forests provide (Pereira 2006).

The interactions between the forest area and the problem of climate change must be analyzed from two perspectives. On the one hand, it is necessary to contemplate what the forests can bring to the reduction of this problem, with a view to mitigation, and on the other hand, what impact climate change can have on forests, with a view to their adaptation and evolution (Torres Ribeiro and Freitas 2010).

These interactions are not independent and are affected by complex interconnection and cause-effect processes. For example, the importance of forests for the mitigation of CO_2 concentrations may be affected if the impact of climatic changes reduces their storage capacity through growth and development constraints or if the problem of forest fires is increased. In other words, it is necessary to define and apply tools to manage forests more effectively, in order to face the problem of climate change, interlinking adaptation and mitigation, with a view to adaptation to mitigate the effects and consequences (Boegelsack et al. 2018; De Groot et al. 2013).

In this book it is intended to approach climate change, starting from the global perspective, where some situations corresponding to environmental impacts provoked or potentiated by climate change are presented, and then to deal specifically with the Portuguese situation from the point of view of impacts.

Next section deals specifically with the evolution of the climate in Portugal, with the intention of understanding how these variations can influence the development of the forest and its species. In this way, an analysis of the current state and the development of the climate in the last years will be presented in the following sections, both in terms of the evolution of the average air temperature and in the slope of the precipitation evolution.

An analysis will also be made of the effects of the changes and their relationship with the occurrence of forest fires in Portugal, with particular attention to the period corresponding to the period 2001–2017, and to the way in which these occurrences interfere in the development and evolution of the forest.

Finally, an analysis is made of the impacts of climate change and its consequences on the Portuguese forest, in particular due to the increase of occurrences of forest fires, pests and dispersal of invasive species.

References

Boegelsack N, Withey J, O'Sullivan G, McMartin D (2018) A critical examination of the relationship between wildfires and climate change with consideration of the human impact. J Environ Prot 9(05):461

Collins (s.d.) Collins English Dictionary. Available https://www.collinsdictionary.com/dictionary/english/climate

Crate SA, Nuttall M (2016) Anthropology and climate change: from encounters to actions. Routledge

De Groot WJ, Flannigan MD, Stocks BJ (2013) Climate change and wildfires. In: González-Cabán A (tech. coord.) Proceedings of the fourth international symposium on fire economics, planning, and policy: climate change and wildfires. General Technical Reports PSW-GTR-245 (English). US Department of Agriculture, Forest Service, Pacific Southwest Research Station, Albany, CA, vol 245, pp 1–10

Dunlap RE, McCright AM (2010) 14 Climate change denial: sources, actors and strategies. In: Routledge handbook of climate change and society, p 240

Füssel H-M (2007) Vulnerability: a generally applicable conceptual framework for climate change research. Glob Environ Change 17(2):155–167

Henriques AG (2009) Convenção quadro das nações unidas sobre alterações climáticas. Instituto Superior Técnico, Mestrado em Engenharia do Ambiente, Lisboa. https://fenix.ist.utl.pt/disciplinas/pa5/2008-2009/2-semestre/convencao-quadro-das-nacoes--unidas-sobre-alteracoes-climaticas

I. P. O. C. Change (2014) "IPCC," Climate change

I. P. O. C. Change (2015) Climate change 2014: mitigation of climate change. Cambridge University Press

Kurten B (2017) Pleistocene mammals of Europe. Routledge

Melillo JM, Richmond T, Yohe G (2014) Climate change impacts in the United States. In: Third national climate assessment, vol 52

Milheiro L (2006) Alterações climáticas em Portugal: cenários, impactos e medidas de adaptação: projecto SIAM II

Mitchum G, Dutton A, Chambers DP, Wdowinski S (2017) Sea level rise. In: Florida's climate: changes, variations, & impacts

Moss RH et al (2010) The next generation of scenarios for climate change research and assessment. Nature 463(7282):747

NOAA (s.d., 25/08/2018) Atmospheric carbon dioxide concentrations and global annual average temperatures over the years 1880 to 2009. Available https://ja.wikipedia.org/wiki/%E3%83%95%E3%82%A1%E3%82%A4%E3%83%AB:Atmospheric_carbon_dioxide_concentrations_and_global_annual_average_temperatures_over_the_years_1880_to_2009.png

Parmesan C, Yohe G (2003) A globally coherent fingerprint of climate change impacts across natural systems. Nature 421(6918):37

Pereira JS et al (2006) Florestas e biodiversidade, Alterações climáticas em Portugal—cenários, impactos e medidas de adaptação (Projecto SIAM II). Gradiva, Lisbon, pp 301–343

Petit JR (s.d.) Climate change: how do we know? In: Nasa Climate Change (ed)

Stocker T (2014) Climate change 2013: the physical science basis: working group I contribution to the fifth assessment report of the intergovernmental Panel on Climate Change. Cambridge University Press

Tian H et al (2016) The terrestrial biosphere as a net source of greenhouse gases to the atmosphere. Nature 531(7593):225

Torres Ribeiro K, Freitas L (2010) Impactos potenciais das alterações no Código Florestal sobre a vegetação de campos rupestres e campos de altitude. Biota Neotropica 10(4)

Varner (2009, 27/08/2018) Pleistocene epoch of the quaternary period. Available http://www.americanroads.us/oceanlinks/pleistocene_europe_map.jpg

Watts N et al (2015) Health and climate change: policy responses to protect public health. Lancet 386(10006):1861–1914

Wikipedia (s.d., 25/08/2018) Timeline of glaciation. Available https://en.wikipedia.org/wiki/Timeline_of_glaciation

World Health Organization (2015) Global technical strategy for malaria 2016–2030. World Health Organization

Zagwijn WH (1989) Vegetation and climate during warmer intervals in the Late Pleistocene of western and central Europe. Quatern Int 3:57–67

Chapter 2
Global Climate Change Outlook

Abstract Climate change is a reality that affects the daily lives of communities around the world, mainly due to the increasingly frequent occurrence of extreme weather phenomena. However, there are other concerns caused by climate change, notably those caused by crop and forest crop growth cycles, and which will be addressed in a separate chapter. Thus, in this chapter the theme is framed, highlighting some examples of situations that occurred on a large scale and all over the world, related to profound changes in climate, with anthropic origin.

Keywords Earth systems · Anthropic actions · Case studies · Climate evolution

Climate changes and human influence have led to the decrease or disappearance of several lakes, rivers and other bodies of water. This does not only affect the environment, but also the communities that support fisheries and agriculture (Crate and Nuttall 2016).

Temperatures on Earth are suitable for life thanks to a natural process called "greenhouse effect". When the solar radiation reaches the atmosphere, part of it is reflected into space, and part of it passes and is absorbed by the Earth. This causes the Earth's surface to warm up. Heat is radiated outward and absorbed by the gases present in the Earth's atmosphere, the so-called "greenhouse gases" (GHG) (Fig. 2.1) (Raval and Ramanathan 1989).

This process prevents heat from disappearing, causing the temperature to rise to +14 °C instead of –19 °C. There are many greenhouse gases responsible for additional warming of the atmosphere, which are produced in different manners. Most come from the combustion of fossil fuels in cars, factories and the production of electricity. The gas responsible for most of the heating is carbon dioxide. Other gases that contribute to heating are methane, expelled by landfills and agriculture (especially the digestive systems of large animals in intensive production), nitrous oxide from fertilizers, gases used for cooling in industrial processes and the massive loss of forest area, otherwise they would store CO_2 (Rodhe 1990; Rosenzweig and Hillel 1998).

Other examples of large-scale environmental impacts caused by the direct action of man, which in a time-prolonged way, can significantly be found to affect the living conditions of populations. These situations, if they initially have influence on

© The Author(s), under exclusive license to Springer Nature Switzerland AG 2020   9
L. J. R. Nunes et al., *Climate Change Impact on Environmental
Variability in the Forest*, SpringerBriefs in Environmental Science,
https://doi.org/10.1007/978-3-030-34417-7_2

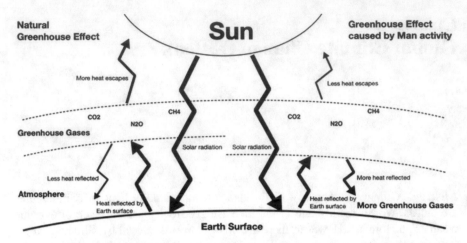

Fig. 2.1 The anthropogenic influence on the greenhouse effect. Adapted from UNP (2018)

a local or regional scale, over time acquire scales of global amplitude, extending their influence in a broad way. In all cases, anthropogenic action associated with the effects of climate change has devastating and hardly reversible effects.

A well-known and studied example of this anthropic influence can be observed in the Aral Sea (Fig. 2.2), in Central Asia, also known as the fourth largest lake in the world, since it has suffered environmental impacts since the 60s of last century. Millions of years ago, the northwestern part of Uzbekistan and southern Kazakhstan were covered by a huge inland sea. When the waters receded, they left a wide plain of highly saline soil. One of the remnants of this ancient sea is the Aral Sea, which is an inland salt water sea, fed by the rivers Amu Darya and Syr Darya. The fresh water of these two rivers kept the water and salt levels of the Aral in perfect balance (Jin et al. 2017).

In the early 1960s, the Soviet government devised a plan for the Soviet Union to become self-sufficient in cotton and to promote rice production. For this reason, the government ordered that the necessary water be withdrawn from the two rivers and that dams be built to feed an 850-km canal with a system of subsidiary long-range channels. When the irrigation system was completed, millions of acres along both sides of the main channel were flooded. In the 30 years following the implementation of this plan, the Aral Sea has suffered a marked decrease in water level, its shoreline has receded and its salt content has increased, rendering the marine environment hostile to life (Jin et al. 2017).

With the disappearance of marine life, fishing also suffered. The Soviet scheme was based on the construction of a series of dams in the two rivers to create reservoirs, from which channels would emerge to divert water to the fields. For fields to flourish with such immense areas of monoculture, it was necessary for farmers to use large quantities of pesticides. With intensive irrigation, the salt was attracted to the soil surface, where it was deposited, rendering the crop unviable (Simurgtravel 2018).

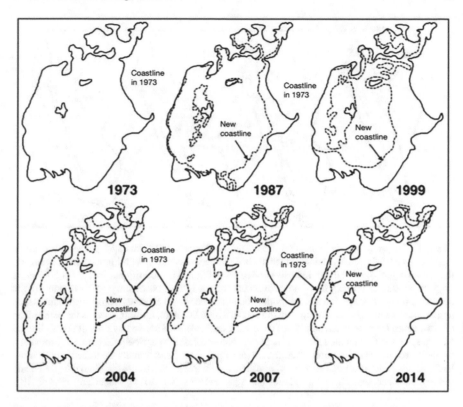

Fig. 2.2 Recent historical evolution of the Aral Sea. Adapted from Simurgtravel (2018)

But it is a phenomenon that is not limited to this concrete case and is repeated in different parts of the planet, as can be seen in the examples presented here.

The world's fourth largest island, Madagascar, underwent a major process of deforestation caused by man's action due to the uncontrolled exploitation of forest charcoal, which has been heavily influenced by climate change and has led to long periods of drought. Before its discovery the island had more than 300,000 km^2 of forest and currently only counts 50,000 km^2 (Fig. 2.3). In the opinion of several experts, if this rate of forest area reduction continues, in 35 years all species of the island's native fauna and flora shall disappear (Desbureaux and Damania 2018).

Lake Chad, which for a long time was the third largest source of drinking water on the African continent, lost 90% of its surface area over 30 years. In the 1960s it covered more than 25,000 km^2 of area, while at present it does not exceed 1000 km^2. In addition, the northern part completely dried and its average depth decreased from seven to two meters (Fig. 2.4).

The lake has undergone several modifications over the years since on previous occasions it was thought that it was going to disappear completely, but it regained part of its size again. Currently due to its shallow depth, it is vulnerable to the variations of the climate that cause the evaporation of its waters (Hansen 2017).

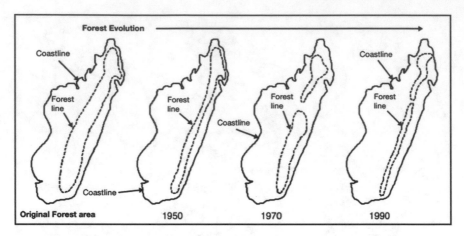

Fig. 2.3 The process of deforestation in Madagascar began a long time ago, practically since its discovery, and has even accelerated since the late nineteenth century, with French colonization and the conversion of forest land to intensive coffee production. The country has lost about 80% of its original forests and primary forest now covers only about 12% of the country. Other reasons for the country's sharp deforestation are the production of firewood and charcoal, forest fires, subsistence agriculture for cutting and burning, grazing and intensive livestock farming. Tavy is the name given to subsistence farming that results from the cutting of the forest followed by fires. It is a traditional Malagasy cultivation method that transforms the native forest into agricultural land. Unfortunately, the newly acquired land can only be cultivated for a limited time and farmers are forced to clear more land and repeat the process. However, where there are no trees, rain cannot seep into the ground and run to the surface, causing sharp soil erosion. Adapted from EOI (2014), MadaCamp.com (2018)

The Dead Sea is another natural resource that has also been greatly affected by human intervention associated with the effects of climate change, particularly due to prolonged drought periods in the region (Fig. 2.5). Its water level is decreasing by one meter every year and so far, it has lost 30% of its size. For this reason, in some areas, the coast is 600 m from the position where it was 20 years ago. The lack of water in the Jordan River is the main cause of this problem, as Israel and Jordan use large quantities of this water to irrigate agricultural crops and domestic consumption (Kiro et al. 2017; Stocker 2013).

According to the European Space Agency (ESA), the ice cap of northern Alaskan lakes has reduced by 22% in 20 years. Between 1991 and 2011 this decrease was evident, as a consequence of climate change, as European scientists explain. During the study years, the layer decreased between 21 and 30 cm and the strongest impact occurred in 2011. These changes affected the availability of water during the winter months. Continuing in this way, the melting of the Arctic, according to the organization Greenpeace, will cause the sea to invade the coastal zone of several parts of the globe (Engram et al. 2018).

Also, the United States National Oceanic and Atmospheric Administration (NOAA) warns of the increased risk of devastating floods if worst case scenarios occur. Its experts say that rising water levels are inevitable and a global problem of

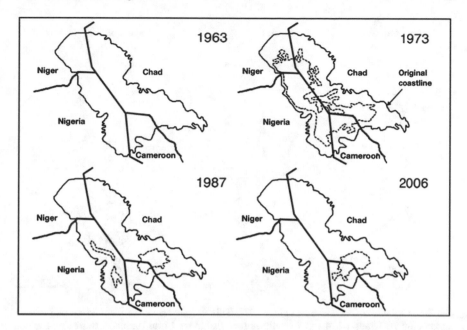

Fig. 2.4 From 1963 until today, more than 90% of lake waters have disappeared, largely due to changes in climatic conditions and increasing human demand for irrigation agriculture, livestock farming, artisanal fishing, and power generation. Images of its progressive decrease (from about 25,000 km^2 in the 60s to less than 1350 km^2 today); of lakeside communities whose sources of subsistence are in decline; droughts, deprivations and conflicts dominate the specialized understanding of the current conditions in the Lake Chad region. These conditions, along with institutional weaknesses and government failures, are the main drivers of vulnerability and insurgency in the region. The conditions at Lake Chad provide a dynamic test field of wide relevance for understanding the major connections between humans and the environment, particularly in relation to sustainability and resilience issues in the Anthropocene. Sustainability, in this sense, encompasses the need to recharge the lake and restore it to a level where it can continue to sustain local livelihoods and regional economies. Where this is not possible, it will be necessary to proactively update the resilience of sites. This can be done by increasing the capacity of lakeside communities to create alternative livelihood options and seek strategies that increase their competence to address the emerging threats associated with the degradation of Lake Chad. Adapted from Anthropocene-curriculum.org (2018)

public knowledge. But this increase may be more severe than predicted (Kuser Olsen 2018).

The fact is that Portugal is not free of this threat either. The country's vulnerability is evident because it is a country with an extensive coastline. For this reason, it requires some concern in its coastal policy, which should take into account future (Ventura et al. 2017).

In an article by RTP on January 25, 2017, Filipe Duarte Santos, Professor of the Faculty of Sciences of the University of Lisbon and researcher at the Center for Ecology, Evolution and Environmental Change (cE3c), stated that "in the 20th century the average level of sea has risen by about 17 cm, and the average rate of climb is increasing, with a rise of two or three millimeters per year" (RTP 2018).

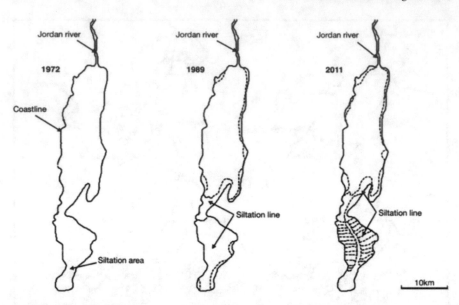

Fig. 2.5 Evolution of the Dead Sea since 1972. This image was drawn from false-color images captured by the Landsat 1, 4 and 7 satellites. It is visible the Lisan Peninsula (lower center) that forms a land bridge across the Dead Sea. Adapted from Keck (2012)

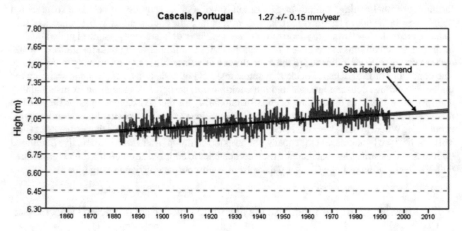

Fig. 2.6 Evolution of the mean sea level in this Portuguese region between 1880 and 2000. The average sea level rise is around 1.27 mm/year with a confidence interval of 95% (±0.15 mm/year) based on the average monthly sea-level data from 1882 to 1993, which is equivalent to a change from 0.42 feet in 100 years. Adapted from (NOAA)

The Cascais mare data, which has been in operation since the 19th century, is presented in this report as one of the evidences that proves that sea level has risen over the last years. The graph presented in Fig. 2.6, provided by NOAA, shows the evolution of mean sea level in this zone of the Portuguese coast between 1880 and 2000.

As can be seen in Fig. 2.6, during this period there is an apparent rise in mean sea level for this region. The novelty is that the pace of sea level rise may be faster than expected. According to the latest studies and observations, notably on the Antarctic ice sheet and its instability, NOAA predicts, in the worst-case scenarios, sea level rise to reach 2.7 m by 2100.

In Spain, the sea could transgress the coastline to 700 m, in lower zones in 2100, due to the effect of climate change. In addition, the increase of one centimeter of the sea level is equivalent to the loss of one meter of beach, as explained by the representatives of the organization (Toimil et al. 2018).

In addition, the IPCC claims that the global average temperature increased by 0.85 degrees between 1880 and 2012, which contributes to more frequent and greater periods of drought, heat waves and habitat degradation around the world (IPOC 2014; IPOC Change 2015; Change 2016).

The IPCC also notes that if the temperature rises to a degree centigrade, ecosystems will face imminent danger, where coral and Arctic reefs will be most affected. This situation is already visible with the coral reef in Belize, as CO_2 concentrations and rising sea temperatures have caused damage to this habitat, which may disappear in 20 years, according to the opinion of several experts (IPOC 2014; Change 2016).

In addition to all of the above, experts point out that rapid population growth and lack of land use planning are the main reasons that will contribute to the reduction of ecosystems, where forest areas are included (Boserup 2017).

Forests, which act as warehouses for greenhouse gases, help mitigate the effects of climate change. However, the biological diversity of forests is also directly and indirectly affected by changing climatic conditions. These changes question the extent to which forests could continue to sequester greenhouse gases in the future (Hui et al. 2017).

Models representing ecosystems and their variations in the different climate scenarios suggest that the changes will present a variety of impacts on the distribution of forest populations, as well as on the impact on the function and composition of ecosystems. In general, habitats are expected to move towards the poles and progress in altitude, conquering new territories (Pecl 2017).

Filipe Duarte Santos, who, in addition to the previously mentioned attributions, is also President of the National Environment Council and led the project "Climate Changes in Portugal: Scenarios, Impacts and Adaptation Measures (SIAM)", in a round table about adaptation to in the context of a National Meeting of Water and Sanitation Management Entities (ENEG 2017) warned that irrigation in Alentejo might not be viable in the medium and long term and suggested that areas further north. The "montado"[1] does not survive, and it will not be diseases, but the lack of water that will extinguish the cork oak in the Alentejo. To maintain the production

---

[1]"Montado" is the Portuguese designation to a forest composed mainly by cork oak and holm oak, very common in South Portugal and Spain. It is a remaining agro-forest-pastoralism system created in the Mediterranean basin.

of cork, it is necessary to help the "montado" to migrate in altitude and to the north, taking into account the places that this year were affected by the fires, "referring to the possibility of this species being used for reforestation of the burned areas" (Clima de Portugal está a ficar como o de Marrocos ou da Tunísia. Accessed on: 25/08/ 2018).

As these habitats change, forest biodiversity will be forced to adapt, and as a result, the composition of species in forests will likely change, and species and populations that are already vulnerable will be potentially outdated. In addition, with climate change, there will be a greater incidence of extreme weather events, such as floods and droughts. These types of events will further affect forest populations and may make forests more prone to disturbances such as fires, invasive species proliferation, diseases and pests (Dale et al. 2001).

A mixed and preferably autochthonous forest stand, consisting of several different tree species with different ecological requirements and the ability to adapt to the expected changes in temperature, precipitation, frequency of storms and pests, will allow for the continuous adjustments according to the climatic evolution (De Dios et al. 2007).

References

Anthropocene-curriculum.org. (s.d., 29/08/2018) Too little water: The Lake Chad Story. Available: https://www.anthropocene-curriculum.org/pages/root/campus-2014/filtering-the-anthropocene/lake-chad-sharing-diminishing-resource/too-little-water-the-lake-chad-story/

Boserup E (2017) The conditions of agricultural growth: the economics of agrarian change under population pressure. Routledge, Abingdon

Change C (2016) What climate change

Crate SA, Nuttall M (2016) Anthropology and climate change: from encounters to actions. Routledge, Abingdon

Dale VH et al (2001) Climate change and forest disturbances: climate change can affect forests by altering the frequency, intensity, duration, and timing of fire, drought, introduced species, insect and pathogen outbreaks, hurricanes, windstorms, ice storms, or landslides. AIBS Bull 51(9):723–734

De Dios VR, Fischer C, Colinas C (2007) Climate change effects on Mediterranean forests and preventive measures. New Forests 33(1):29–40

Desbureaux S, Damania R (2018) Rain, forests and farmers: evidence of drought induced deforestation in Madagascar and its consequences for biodiversity conservation. Biol Cons 221:357–364

Engram M, Arp CD, Jones BM, Ajadi OA, Meyer FJ (2018) Analyzing floating and bedfast lake ice regimes across Arctic Alaska using 25 years of space-borne SAR imagery. Remote Sens Environ 209:660–676

EOI (s.d.). Deforestation in Madagascar: a threat to its biodiversity. Available: http://www.eoi.es/blogs/guidopreti/2014/02/04/deforestation-in-madagascar-a-threat-to-its-biodiversity/

Hansen K (2017) The rise and fall of Africa's Great Lake: scientists try to understand the fluctuations of Lake Chad: feature articles

Hui D, Deng Q, Tian H, Luo Y (2017) Climate change and carbon sequestration in forest ecosystems. In: Handbook of climate change mitigation and adaptation, pp 555–594

IPOC (2014) Change, "IPCC," Climate change

IPOC Change (2015) Climate change 2014: mitigation of climate change. Cambridge University Press, Cambridge

Jin Q, Wei J, Yang Z-L, Lin P (2017) Irrigation-induced environmental changes around the Aral Sea: an integrated view from multiple satellite observations. Remote Sens 9(9):900

JN, Clima de Portugal está a ficar como o de Marrocos ou da Tunísia. Accessed on: 25/08/2018. Available: https://www.jn.pt/nacional/interior/clima-de-portugal-esta-a-ficar-como-o-de-marrocos-ou-da-tunisia-8933444.html

Keck A (2012, 29/08/2018). NASA sees new salt in an ancient sea. Available: https://phys.org/news/2012-04-nasa-salt-ancient-sea.html

Kiro Y et al (2017) Relationships between lake-level changes and water and salt budgets in the Dead Sea during extreme aridities in the Eastern Mediterranean. Earth Planet Sci Lett 464:211–226

Kuser Olsen V et al (2018) An approach for improving flood risk communication using realistic interactive visualisation. J Flood Risk Manage 11:S783–S793

MadaCamp.com. (s.d., 02/09/2018). Tavy agriculture. Available: https://www.madacamp.com/Tavy

NOAA, Mean Sea Level Trend 210–021 Cascais, Portugal, ed. USA: NOAA, s.d.

Pecl GT et al (2017) Biodiversity redistribution under climate change: impacts on ecosystems and human well-being. Science 355(6332):caai9214

Raval A, Ramanathan V (1989) Observational determination of the greenhouse effect. Nature 342(6251):758

Rodhe H (1990) A comparison of the contribution of various gases to the greenhouse effect. Science 248(4960):1217–1219

Rosenzweig C, Hillel D (1998) Climate change and the global harvest: potential impacts of the greenhouse effect on agriculture. Oxford University Press, Oxford

RTP, Litoral português em risco com a subida crescente do nível das águas do mar. Accessed on: 24/08/2018. Available: https://www.rtp.pt/noticias/pais/litoral-portugues-em-risco-com-a-subida-crescente-do-nivel-das-aguas-do-mar_n978444

Simurgtravel (s.d., 27/08/2018). The Aral Sea. Available http://simurgtravel.com/the-aral-sea

Stocker TF et al (2013) Climate change 2013: the physical science basis. Working group i contribution to the fifth assessment report of the intergovernmental panel on climate change 2013. https://www.google.com/books

Toimil A, Díaz-Simal P, Losada IJ, Camus P (2018) Estimating the risk of loss of beach recreation value under climate change. Tour Manag 68:387–400

U. N. P. Service. (s.d., 27/08/2018). Efecto Invernadero inducido por el ser humano. Available https://www.ekoenergy.org/es/extras/background-information/climate-change/

Ventura C, Sousa J, Fernandes A (2017) Os estuários e as alterações climáticas: impactes da subida do nível médio das águas do mar em Vila Franca de Xira. GOT, Revista de Geografia e Ordenamento do Território 11:327–350

Chapter 3
The Evolution of Climate Change in Portugal

Abstract As a phenomenon affecting countries around the entire world, those located in regions with a predominantly Mediterranean climate have continuously suffered the effects of climate change. Portugal, being a country that is located in the area of influence of Mediterranean climate, has also felt these variations, which highlight the periods of drought even during winter. This chapter presents a characterization of the climate in mainland Portugal. Also, noteworthy is the analysis of the evolution of average monthly air temperatures from 2000 to 2017, using the same procedure for precipitation over the same period of time. Thus, it was possible to determine the existence of a trend for the occurrence of climate anomalies, which influence, for example, the increased risk for the occurrence of more rural fires.

Keywords Mediterranean climate · Climate anomalies · Monthly average air temperature · Monthly average precipitation

3.1 Framework

The problem of climate change has been continuously addressed in Portugal, both by the elements of the academic environment, which have been warning of the causes and consequences of the phenomenon, but also by other sectors of civil society, including political and which has for some time initiated an ambitious program for the implementation of measures aimed at minimizing the negative impacts of climate change in the country.

Although the first steps are already being taken to effectively understand the causes and consequences, it is assumed by the national and international scientific community that countries with Mediterranean climate characteristics may be most affected by climate change (Pereira 2006; Giorgi and Lionello 2008; Pausas 2004).

In this sense, a significant effort has been made in Portugal to implement measures that contribute to mitigating the harmful effects of climate change, which at least begin to have international recognition.

© The Author(s), under exclusive license to Springer Nature Switzerland AG 2020
L. J. R. Nunes et al., *Climate Change Impact on Environmental Variability in the Forest*, SpringerBriefs in Environmental Science,
https://doi.org/10.1007/978-3-030-34417-7_3

On June 18, 2018, TSF radio, on its website, published a news item entitled "Portugal is second in a ranking on ambition in goals and measures to comply with the Paris Agreement against climate change, being only exceeded by Sweden" (TSF/LUSA 2018).

"The vast majority of the Member States of the European Union (EU) are failing to meet the targets of the Paris Agreement, and Portugal is among the few countries that have appealed to targets and policies more ambitious in the area of energy and climate, such as reducing greenhouse gas emissions" (TSF/LUSA 2018).

This information is the result of a study entitled "Off target: Ranking of EU countries' ambition and progress in fighting climate change", which determines the Member States' commitment to achieving energy policy and energy targets and climate change, and what progress they are making in reducing greenhouse gas emissions and implementing programs for the use of renewable energy and energy efficiency (TSF/LUSA 2018).

As previously mentioned, climate can be defined as the set of long-term meteorological conditions prevailing in a given area. Thus, in this perspective, the average values of the climatic variables of a given location will be more representative according to the time interval used in the analysis, thus constituting a time series.

In this way, the same results are not obtained when comparing a time interval of one year with one of ten years, or one of a hundred years. It is important to have long time series of data to analyze the variations and the evolution of the climate. For example, the Portuguese Institute of the Sea and the Atmosphere (IPMA), has available series of meteorological data dating back to 1865 (IPMA 2018).

The World Meteorological Organization (WMO) has agreed that the characterization of climate is done by analyzing the mean values of the various climatic elements over a period of 30 years. This period is the normal value of a climatic element and represents the mean value corresponding to a sufficiently long number of years to be assumed to represent the predominant value of that element at the site under consideration (IPMA 2018).

Likewise, the WMO designates the statistical values obtained for periods of 30 years by climatological norm, starting this period in the first year of each set of years under analysis (1901–30, 1931–1960, 1961–1990, 1991–2020). These are the normals of reference, however, they can be calculated and used normal climatologicals based on intercalary periods, for example periods of the genre 1951–80 or 1971–2000 (IPMA 2018).

The Portuguese Institute of the Sea and Atmosphere (IPMA) provides online information on the climatological norm of 21 meteorological stations for the period 1971–2000, including monthly and annual values of the main climatic elements. In the same website are also available the average values of air temperature and total precipitation (IPMA 2018).

Based on the results of the climatological norm of the period 1971–2000, it is possible to identify the different climate types for Continental Portugal, using the Köppen-Geiger classification (Fig. 3.1) (IPMA 2018; Peel et al. 2007).

In order to assess the current state of the climate in Portugal and to verify how climate change has occurred in the following sections, namely in Sect. 3.2. The

Fig. 3.1 According to this climatic classification, most of the continental territory in a Temperate climate of Type C, being verified the Subtype Cs (Temperate with dry Summer) and with the following varieties: Csa, temperate climate with hot summer and in the interior regions of the Douro valley (part of the district of Bragança), as well as in the southern regions of the Montejunto-Estrela mountain system (except in the west coast of the Alentejo and Algarve); Csb, temperate climate with dry and mild summer, in almost all regions to the north of the Montejunto-Estrela mountain system and in the regions of the west coast of Alentejo and Algarve; and in a restricted region of the Lower Alentejo, in the district of Beja, one can find Arid Climate—Type B, Subtype BS (steppe climate), BSk variety (medium latitude cold steppe climate). Adapted from IPMA (2018)

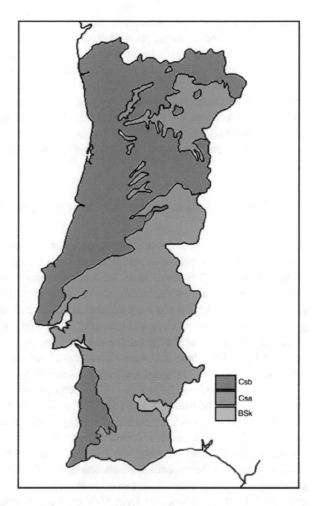

evolution of mean air temperature, and in Sect. 3.3. The evolution of the precipitation, will be presented an analysis to the data available on the website of the IPMA (www. ipma.pt).

For the two climatic parameters selected, mean air temperature and precipitation, the IPMA provides information on the occurrence of anomalies by creating maps of isolines representing, in the case of average air temperature, areas where the air temperature exceeded (positive or negative) values of the climatic norm (period from 1971 to 2000). Likewise, for precipitation, isoline maps are also presented, thus representing the percentage of precipitation compared to the climatic normal (1971–2000).

Subsequently, new maps of isolines were constructed based on the previous ones, made available by the IPMA website, which were simplified in order to facilitate the reading and counting of the number of anomalies that occurred, both for the average temperature of the air as for precipitation.

After counting the anomalies between 2001 and 2017 for the two parameters selected, a table with the presentation of the data was elaborated. In order to verify the existence of a trend of occurrence of events, a moving average model was applied. In this way, it is possible to relate the data obtained with the information previously presented with the impacts caused by climate change, which addresses the issue of the occurrence of forest fires, in order to relate their intensity and severity to the climatic anomalies recorded during the study period.

3.2   The Evolution of the Air Average Temperature

The best-known and referred parameter when addressing the subject of climate change is surely the rise in air temperature. If on the one hand the scientific community addresses other parameters with the same concern and capacity for analysis, civil society refers to this particular subject, often without understanding its real effects, but mainly because it is the most addressed by the media.

It is a rare day when no news comes out in all kinds of media that do not allude to "global warming" and the "greenhouse effect", or very specifically to its effects and consequences anywhere in the world, such as occurrences of forest fires, hurricanes, floods, long periods of drought, sea level rise or changes in monsoon cycles.

Thus, in this section the analysis of the occurrence of anomalies in the average air temperature will be made. In this particular case, the occurrence of an anomaly will be considered whenever the average monthly temperature exceeds 1 °C, compared to normal climatic conditions. Based on this analysis, isoline maps were constructed, which are presented in the Appendix 1 in Fig. 3.9a and below, which visually indicate which regions of the country have exceeded +1 °C (colored in red), −1 °C (colored to blue) and which were similar to climatic norms (colored to green). An abnormal month is considered when at least 50% of the national territory has been subject to temperature values above or below 1 °C compared to normal climatic conditions (Fig. 3.2).

Figure 3.3 shows the anomalies observed in the period between 2001 and 2017. As can be seen, there was a significant set of anomalies in all the constituent years of the period, with a maximum of anomalies of 7 occurring in 2017, 2005, 2006, 2009, 2015 and 2017. The lowest number of anomalies was reached in 2004 and 2007, with 3 occurrences.

Based on these data, it is necessary to determine the existence of a tendency for an increasing number of anomalies, that is, to verify if the number of anomalies occurring in the period between 2001 and 2017 presents a tendency of occurrence in some sense. For this inquiry the Simple Moving Average Method was used (Crowder 1987).

The determination of the simple moving average of a set of n elements is obtained by calculating the unweighted averages of the subsets of n elements in a given set

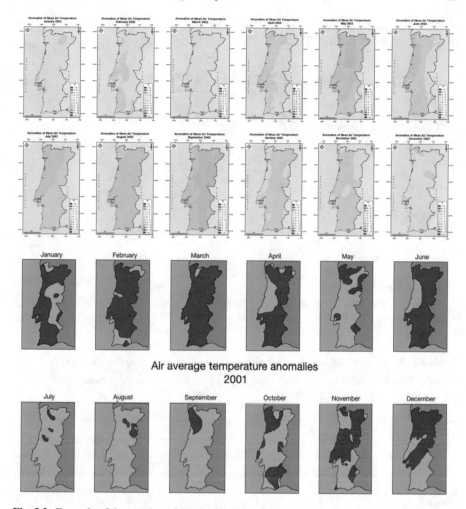

Fig. 3.2 Example of the creation of maps of isolines for the anomalies of year 2001. In the upper part are the images provided by the IPMA and in the bottom are the images created from the first. Adapted from IPMA (2018)

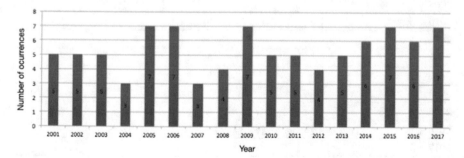

Fig. 3.3 Count of anomalies verified in the period between 2001 and 2017

of data. For example, given a set of n elements $p_1, \ldots, p_n$, the first element of the moving average is given by Eq. 3.1:

$$Mn = \frac{p_1 + \cdots + p_n}{n} = \frac{1}{n} \sum_{i=1}^{n} p_i \qquad (3.1)$$

The second is given by Eq. 3.2:

$$Mn' = \frac{p_2 + \cdots + p_{n+1}}{n} = \frac{1}{n} \sum_{i=2}^{n+1} p_i \qquad (3.2)$$

Or even by Eq. 3.3:

$$Mn' = Mn + \frac{p_{n+1}}{n} - \frac{p_1}{n} \qquad (3.3)$$

and so on until $p_{n-n+1}, \ldots, p_n$.

Table 3.1 presents the results of the application of the Simple Moving Average Method and in Fig. 3.4 the trend lines are presented. In this particular situation the

Table 3.1 Results obtained from the application of the simple moving average method

Year	No. of occurrences	Average of 3 periods	Average of 5 periods	Average of 7 periods
2001	5			
2002	5	5		
2003	5	4	5	
2004	3	5	5	5
2005	7	6	5	5
2006	7	6	5	5
2007	3	5	6	5
2008	4	5	5	5
2009	7	5	5	5
2010	5	6	5	5
2011	5	5	5	5
2012	4	5	5	6
2013	5	5	5	5
2014	6	6	6	6
2015	7	6	6	
2016	6	7		
2017	7			

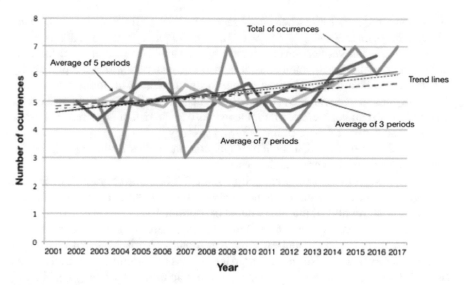

Fig. 3.4   Trend lines

method was applied for 3, 5 and 7 periods, since it was understood that for the number of data available, it would be sufficient.

The application of the periods consists in the aggregation of groups of data according to the period used, and in the realization of the average of this group, and the result obtained corresponding to the average element of the group.

For example, for an average of 3 periods, the first 3 years, 2001, 2002 and 2003 are selected, and the average number of occurrences for these three years is determined. The final result is assigned to the middle element, in this case it is the year 2002. Then the operation was repeated for the group of 3 following elements, 2002, 2003 and 2004, and so on.

For the calculation of the means of 5 and 7 periods, the procedure is similar, but now, instead of selecting 3 elements, 5 or 7 will be selected, repeating the operation for all the elements of the sample. The higher the number of data available, the more means with different periods can be calculated, the data being analyzed as linearly as possible. It is in this way that one can determine if there is a growing or decreasing tendency of the occurrence of a given event.

In the previous figure the projection of the data previously presented in Table 3.1 is shown. As can be seen, the real data or the total occurrences, represented by the blue line, allow the creation of a trend line and indicate by itself a perspective increase in the number of occurrences.

However, with the application of the simple moving averages method, it is verified that the lines corresponding to each of the averages, the average of 3 periods being represented by the red line, the average of the 5 periods represented by the green line and the average of 7 periods represented by the purple line, indicate in a much more visible way an increasing tendency for the occurrence of climatic anomalies. This fact is even more noticeable when one observes the lines or lines of tendency, which clearly indicate an increasing trend towards a more frequent occurrence of this event.

In addition to this verification of the tendency for a given event to occur, in this case the occurrence of climatic anomalies, it is also necessary to verify the probability of occurrences occurring in a given month to the detriment of another, since this seasonality associated with the occurrence of precipitation and the biological cycles of plant growth can be determinant for the increase in the risk of occurrence of forest fires and also for their degree of intensity and severity.

Table 3.2 presents the distribution of climatic anomalies by the months of the constituent years of the period under analysis.

As can be seen in the data presented in Table 3.2 and corroborated by Fig. 3.5, although there is a dispersion for all the months of the year, since there were climatic anomalies in all months of the year without exception.

Over the last 17 years under analysis in this study, there is a greater concentration of these events in the spring and early summer months (April, May and June), so if there are factors that are also anomalous in the months and summer and autumn, also associated with the occurrence of anomalies in precipitation, a strong probability of occurrence of ideal conditions for the outbreak of forest fires of great intensity and severity.

As can be seen, there is also a strong tendency for air temperature anomalies to occur during the summer and autumn months, so it can be said that the probability of repeating situations such as those occurring in 2017, with very favorable conditions for the outbreak of fires in summer and autumn is very high.

3.3   The Evolution of the Precipitation

Similar to the methodology used for the previous section, maps of isolines were created on the maps made available on the IPMA website (www.ipma.pt). In these maps, which are presented in the Appendix 2 in Fig. 3.26a et seq. The zones of the country where precipitation was at least equal to that of the normal climatic period used in this study, the period from 1971 to 2000 were defined. As an anomaly when in a given month, in at least 50% of the continental national territory, the precipitation was lower than that occurred during the period of normal climatic conditions.

Figure 3.6 shows the precipitation anomalies observed between 2001 and 2017. As can be seen, there was a significant set of anomalies in all the constituent years of the period, with a maximum of 11 anomalies occurring until 2017, in the years of 2015 and 2017, but in 2008 there were 10 anomalies, and in the years 2004, 2007, 2009 and 2012, there were 9 anomalies. The lowest number of anomalies was reached in the years 2006 and 2014, with 5 occurrences.

Table 3.3 presents the results of the application of the Simple Moving Average Method and in Fig. 3.7 the trend lines are presented, using the same methodology described in the previous section. Also, in this situation the method was applied for

Table 3.2 Distribution of climatic anomalies by the months of the constituent years of the period under analysis

Year	J	F	M	A	M	J	J	A	S	O	N	D
2001	a	a	a	a		a						
2002	a	a	a	a								a
2003			a		a	a		a	a			
2004	a					a	a					
2005			a	a	a	a	a	a		a		
2006				a	a		a	a	a	a	a	
2007		a		a	a							
2008	a	a		a		a						
2009			a		a	a		a	a	a	a	
2010				a		a	a	a	a	a		
2011				a	a	a			a			
2012			a		a	a			a			
2013	a						a	a	a	a		
2014	a	a			a				a	a	a	
2015				a	a	a	a			a	a	
2016	a					a	a	a	a	a		a
2017		a		a	a	a	a	a		a		
Total	7	6	6	10	10	12	8	8	9	9	4	2

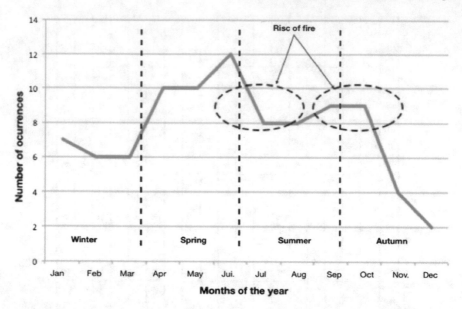

Fig. 3.5 Distribution of climatic anomalies by the months of the constituent years of the period under analysis

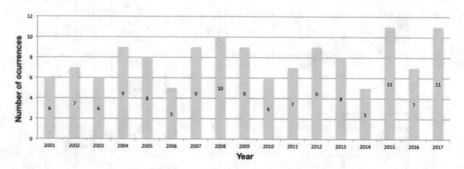

Fig. 3.6 Count of anomalies verified in the period between 2001 and 2017

3, 5 and 7 periods, since it was understood that for the number of data available, it would be sufficient.

Based on these results, we constructed the graph that is presented in Fig. 3.7, where the trend lines are also projected, and the projection of data previously presented in Table 3.3 is shown.

Table 3.3 Results obtained from the application of the simple moving average method

Year	No. of occurrences	Average of 3 periods	Average of 5 periods	Average of 7 periods
2001	6			
2002	7	6		
2003	6	7	7	
2004	9	8	7	7
2005	8	7	7	8
2006	5	7	8	8
2007	9	8	8	8
2008	10	9	8	8
2009	9	8	8	8
2010	6	7	8	8
2011	7	7	8	8
2012	9	8	7	8
2013	8	7	8	8
2014	5	8	8	8
2015	11	8	8	
2016	7	10		
2017	11			

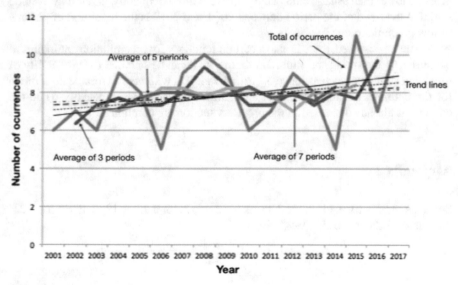

Fig. 3.7 Trend lines

As can be seen, the actual data or total occurrences, represented by the blue line, allow the creation of a trend line and indicate in itself a perspective of growth in the number of occurrences. However, with the application of the simple moving averages method, it is verified that the lines correspond to each of the means, the average of three periods being represented by the red line, the average of the 5 periods represented by the green line and the average of 7 periods represented by the purple line, indicate in a much more visible way an increasing tendency for the occurrence of climatic anomalies.

This fact is even more noticeable when one observes the straight lines or trends, which clearly indicate an increasing trend towards a more frequent occurrence of this event, similar to what had already happened in the previous section, in the analysis of temperature anomalies average of the air.

As can be seen in the data presented in Table 3.4 and corroborated by Fig. 3.8, there is a dispersion of the occurrence of anomalies for all months of the year, since climatic anomalies occurred in all months of the year without exception.

Over the last 17 years under analysis in this study, there is a greater concentration of these events in the spring and early summer months (April, May and June), so if factors are also anomalous in the following months, Summer and autumn, also associated to the occurrence of anomalies in precipitation, a strong probability of occurrence of ideal conditions for the outbreak of forest fires of great intensity and severity.

It also seems evident that the occurrence of periods with or without precipitation at levels lower than usual would occur more and more frequently, and it may even be said that the trend of the period between 2001 and 2017 is the repetition of periods of low precipitation.

As can be observed, there is also a certain tendency for precipitation anomalies to occur during the spring and summer months, so it can be said that the probability of repeating situations such as those occurring in 2017, with very favorable conditions for the outbreak of fires in summer and autumn is very high, especially when the conditions already discussed in the previous section are combined.

Appendix 1

See Figs. 3.9, 3.10, 3.11, 3.12, 3.13, 3.14, 3.15, 3.16, 3.17, 3.18, 3.19, 3.20, 3.21, 3.22, 3.23, 3.24 and 3.25.

Table 3.4 Distribution of climatic anomalies by the months of the constituent years of the period under analysis

Year	J	F	M	A	M	J	J	A	S	O	N	D
2001	a		a		a	a	a				a	a
2002	a	a		a	a	a	a	a				
2003		a		a	a	a	a		a			a
2004	a	a	a	a		a	a	a	a		a	a
2005	a	a		a		a	a		a			a
2006	a	a	a	a	a							a
2007	a	a	a	a	a	a	a			a	a	a
2008	a	a	a			a	a	a	a	a	a	a
2009		a	a	a	a	a	a	a	a	a	a	
2010				a	a	a	a	a	a		a	
2011	a				a	a	a		a	a		a
2012	a	a	a	a	a	a	a	a				a
2013		a		a	a	a	a	a			a	a
2014			a		a		a	a	a			a
2015	a	a	a	a	a	a	a	a	a		a	a
2016			a	a		a	a	a	a	a		a
2017	a	a		a	◼	a	a	a	a	a	a	a
Total	11	12	9	11	12	13	15	11	10	6	9	14

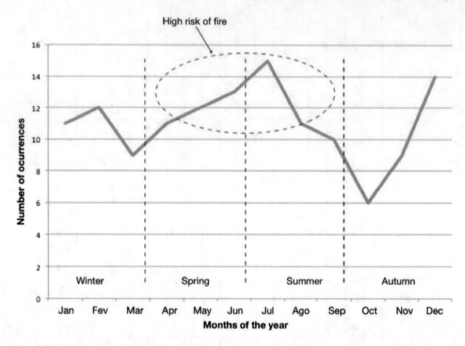

Fig. 3.8 Distribution of climatic anomalies by the months of the constituent years of the period under analysis

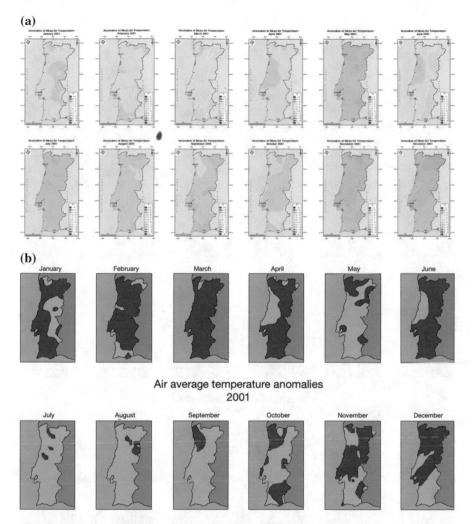

Fig. 3.9 **a** Air temperature anomalies in the year 2001 (IPMA 2018). **b** Representation of regions with average air temperature above, below or equal to the normal climate from 1971 to 2000. The color blue represents the region with average lower air temperature, the color red represents the region where the average air temperature was higher, and the green color represents the region where the average air temperature was equal to the corresponding climate normal period from 1971 to 2000. *Source* Own elaboration and IPMA (2018)

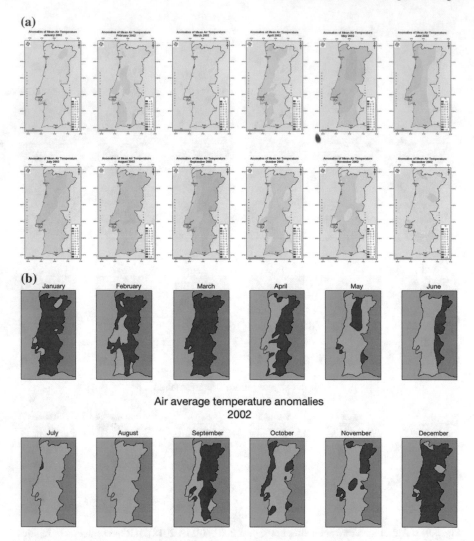

Fig. 3.10 **a** Air temperature anomalies in the year 2002 (IPMA 2018). **b** Representation of regions with average air temperature above, below or equal to the normal climate from 1971 to 2000. The color blue represents the region with average lower air temperature, the color red represents the region where the average air temperature was higher, and the green color represents the region where the average air temperature was equal to the corresponding climate normal period from 1971 to 2000. *Source* Own elaboration and IPMA (2018)

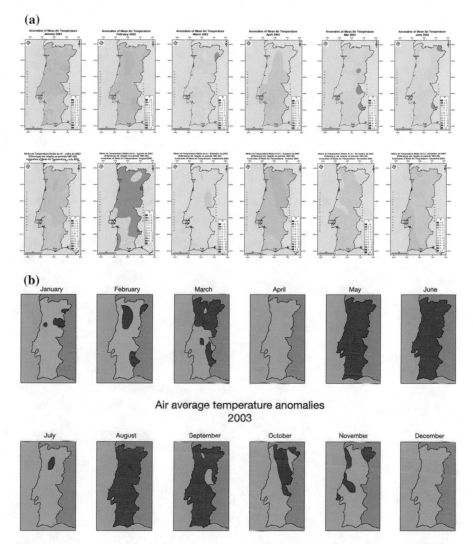

Fig. 3.11 a Air temperature anomalies in the year 2003 (IPMA 2018). **b** Representation of regions with average air temperature above, below or equal to the normal climate from 1971 to 2000. The color blue represents the region with average lower air temperature, the color red represents the region where the average air temperature was higher, and the green color represents the region where the average air temperature was equal to the corresponding climate normal period from 1971 to 2000. *Source* Own elaboration and IPMA (2018)

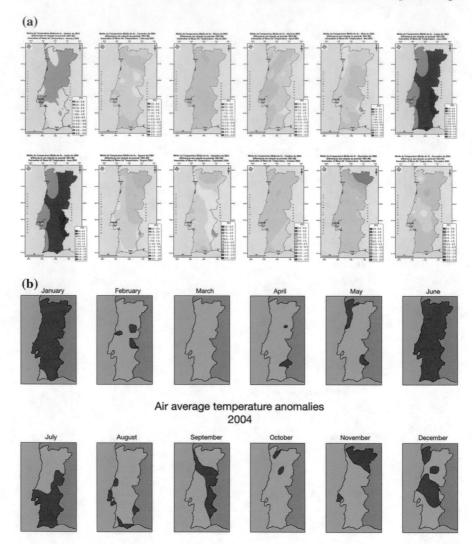

Fig. 3.12 a Air temperature anomalies in the year 2004 (IPMA 2018). **b** Representation of regions with average air temperature above, below or equal to the normal climate from 1971 to 2000. The color blue represents the region with average lower air temperature, the color red represents the region where the average air temperature was higher, and the green color represents the region where the average air temperature was equal to the corresponding climate normal period from 1971 to 2000. *Source* Own elaboration and IPMA (2018)

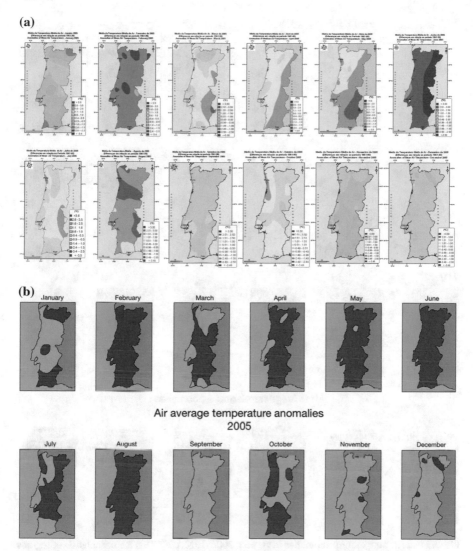

Fig. 3.13 **a** Air temperature anomalies in the year 2005 (IPMA 2018). **b** Representation of regions with average air temperature above, below or equal to the normal climate from 1971 to 2000. The color blue represents the region with average lower air temperature, the color red represents the region where the average air temperature was higher, and the green color represents the region where the average air temperature was equal to the corresponding climate normal period from 1971 to 2000. *Source* Own elaboration and IPMA (2018)

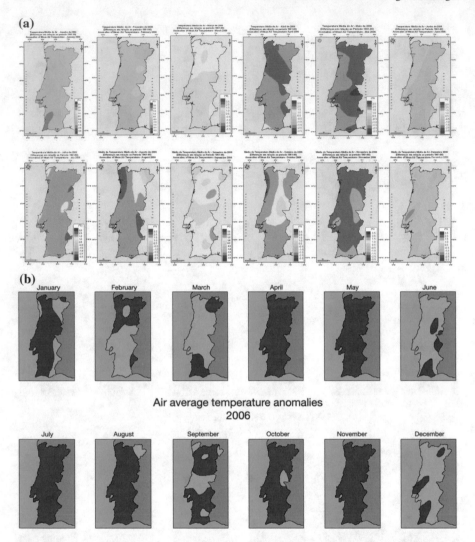

Fig. 3.14 **a** Air temperature anomalies in the year 2006 (IPMA 2018). **b** Representation of regions with average air temperature above, below or equal to the normal climate from 1971 to 2000. The color blue represents the region with average lower air temperature, the color red represents the region where the average air temperature was higher, and green color represents the region where the average air temperature was equal to the corresponding climate normal period from 1971 to 2000. *Source* Own elaboration and IPMA (2018)

(a)

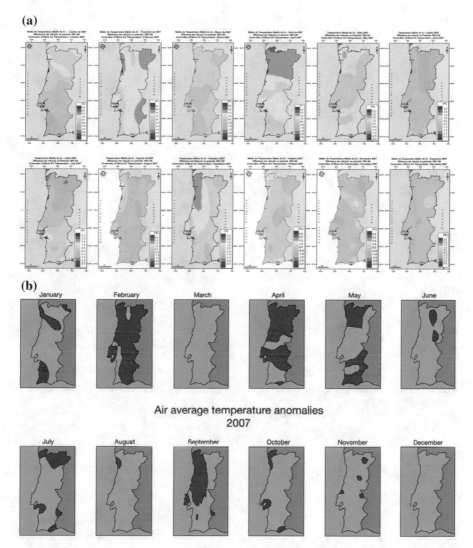

Fig. 3.15 **a** Air temperature anomalies in the year 2007 (IPMA 2018). **b** Representation of regions with average air temperature above, below or equal to the normal climate from 1971 to 2000. The color blue represents the region with average lower air temperature, the color red represents the region where the average air temperature was higher, and the green color represents the region where the average air temperature was equal to the corresponding climate normal period from 1971 to 2000. *Source* Own elaboration and IPMA (2018)

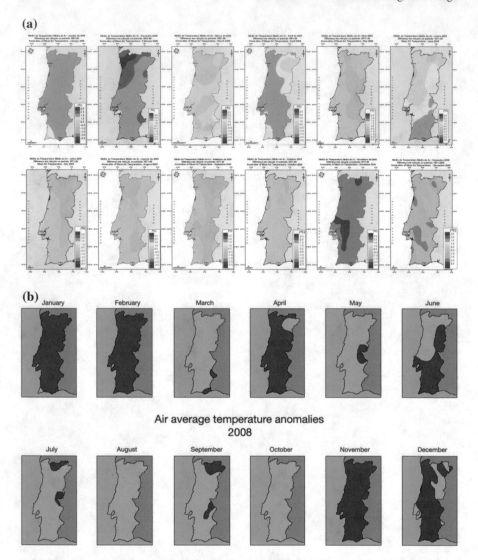

Fig. 3.16 **a** Air temperature anomalies in the year 2008 (IPMA 2018). **b** Representation of regions with average air temperature above, below or equal to the normal climate from 1971 to 2000. The color blue represents the region with average lower air temperature, the color red represents the region where the average air temperature was higher, and the green color represents the region where the average air temperature was equal to the corresponding climate normal period from 1971 to 2000. *Source* Own elaboration and IPMA (2018)

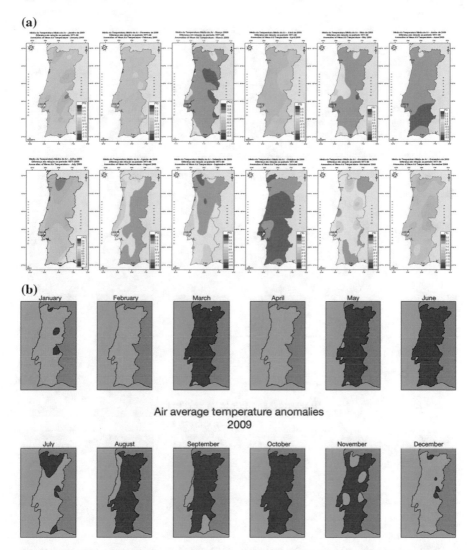

Fig. 3.17 a Air temperature anomalies in the year 2009 (IPMA 2018). **b** Representation of regions with average air temperature above, below or equal to the normal climate from 1971 to 2000. The color blue represents the region with average lower air temperature, the color red represents the region where the average air temperature was higher, and the green color represents the region where the average air temperature was equal to the corresponding climate normal period from 1971 to 2000. *Source* Own elaboration and IPMA (2018)

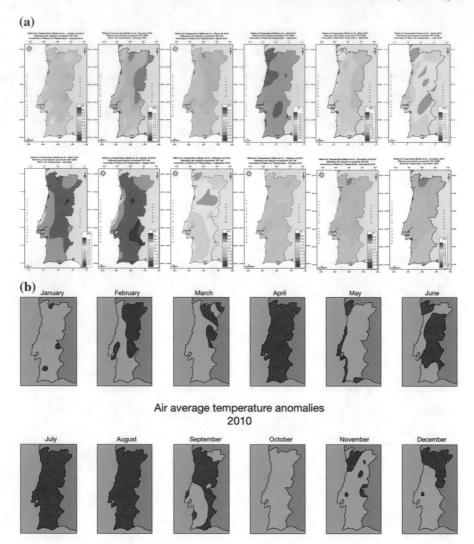

Fig. 3.18 **a** Air temperature anomalies in the year 2010 (IPMA 2018). **b** Representation of regions with average air temperature above, below or equal to the normal climate from 1971 to 2000. The color blue represents the region with average lower air temperature, the color red represents the region where the average air temperature was higher, and the green color represents the region where the average air temperature was equal to the corresponding climate normal period from 1971 to 2000. *Source* Own elaboration and IPMA (2018)

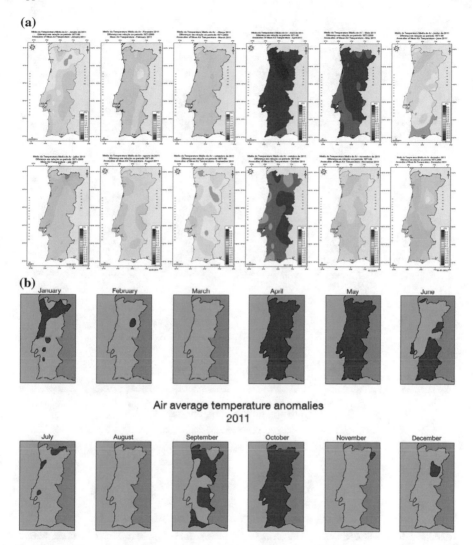

Fig. 3.19 a Air temperature anomalies in the year 2011 (IPMA 2018). **b** Representation of regions with average air temperature above, below or equal to the normal climate from 1971 to 2000. The color blue represents the region with average lower air temperature, the color red represents the region where the average air temperature was higher, and the green color represents the region where the average air temperature was equal to the corresponding climate normal period from 1971 to 2000. *Source* Own elaboration and IPMA (2018)

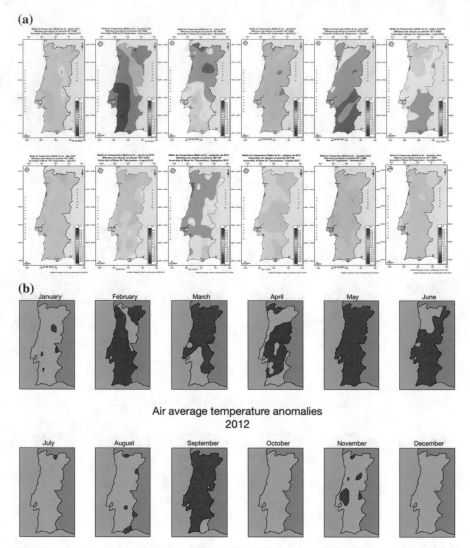

Fig. 3.20 **a** Air temperature anomalies in the year 2012 (IPMA 2018). **b** Representation of regions with average air temperature above, below or equal to the normal climate from 1971 to 2000. The color blue represents the region with average lower air temperature, the color red represents the region where the average air temperature was higher, and the green color represents the region where the average air temperature was equal to the corresponding climate normal period from 1971 to 2000. *Source* Own elaboration and IPMA (2018)

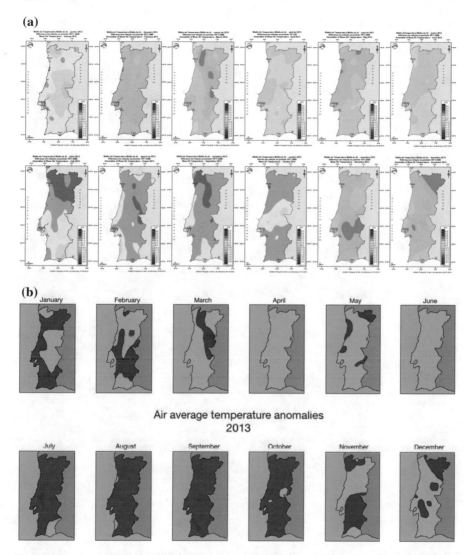

Fig. 3.21 **a** Air temperature anomalies in the year 2013 (IPMA 2018). **b** Representation of regions with average air temperature above, below or equal to the normal climate from 1971 to 2000. The color blue represents the region with average lower air temperature, the color red represents the region where the average air temperature was higher, and the green color represents the region where the average air temperature was equal to the corresponding climate normal period from 1971 to 2000. *Source* Own elaboration and IPMA (2018)

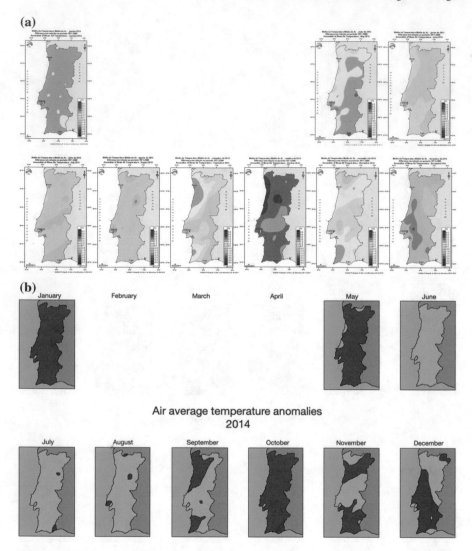

Fig. 3.22   **a** Air temperature anomalies in the year 2014 (IPMA 2018). **b** Representation of regions with average air temperature above, below or equal to the normal climate from 1971 to 2000. The color blue represents the region with average lower air temperature, the color red represents the region where the average air temperature was higher, and the green color represents the region where the average air temperature was equal to the corresponding climate normal period from 1971 to 2000. *Source* Own elaboration and IPMA (2018)

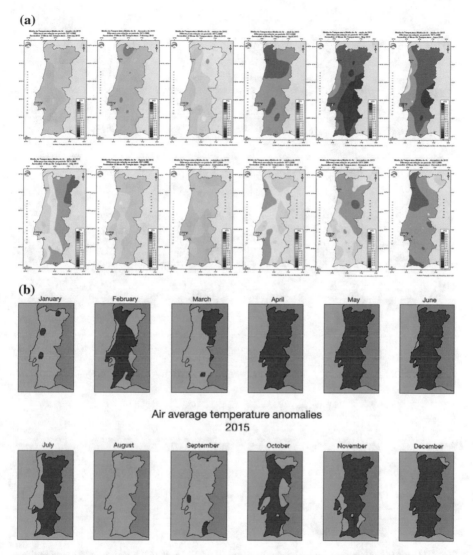

Fig. 3.23 **a** Air temperature anomalies in the year 2015 (IPMA 2018). **b** Representation of regions with average air temperature above, below or equal to the normal climate from 1971 to 2000. The color blue represents the region with average lower air temperature, the color red represents the region where the average air temperature was higher, and the green color represents the region where the average air temperature was equal to the corresponding climate normal period from 1971 to 2000. *Source* Own elaboration and IPMA (2018)

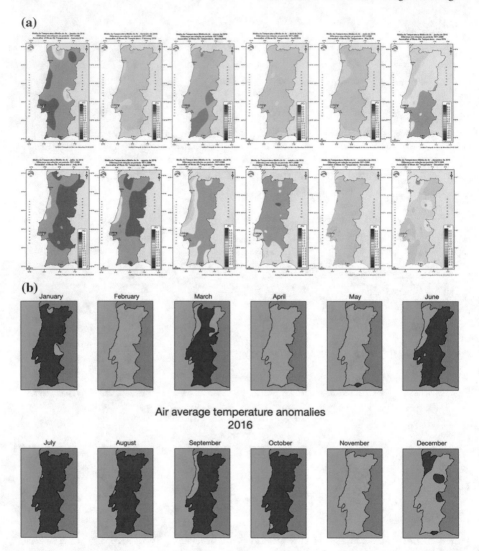

Fig. 3.24 **a** Air temperature anomalies in the year 2016 (IPMA 2018). **b** Representation of regions with average air temperature above, below or equal to the normal climate from 1971 to 2000. The color blue represents the region with average lower air temperature, the color red represents the region where the average air temperature was higher, and the green color represents the region where the average air temperature was equal to the corresponding climate normal period from 1971 to 2000. *Source* Own elaboration and IPMA (2018)

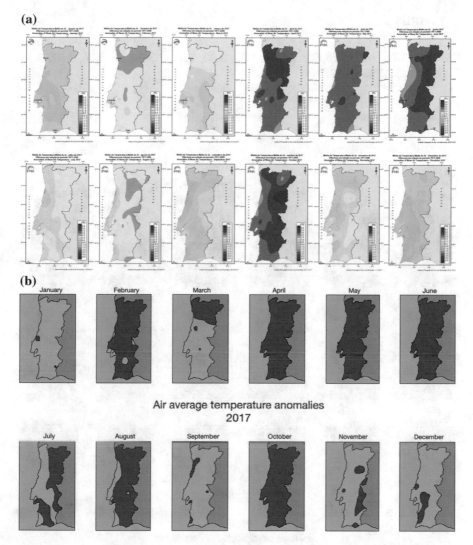

Fig. 3.25 **a** Air temperature anomalies in the year 2017 (IPMA 2018). **b** Representation of regions with average air temperature above, below or equal to the normal climate from 1971 to 2000. The color blue represents the region with average lower air temperature, the color red represents the region where the average air temperature was higher, and the green color represents the region where the average air temperature was equal to the corresponding climate normal period from 1971 to 2000. *Source* Own elaboration and IPMA (2018)

Appendix 2

See Figs. 3.26, 3.27, 3.28, 3.29, 3.30, 3.31, 3.32, 3.33, 3.34, 3.35, 3.36, 3.37, 3.38, 3.39, 3.40, 3.41 and 3.42.

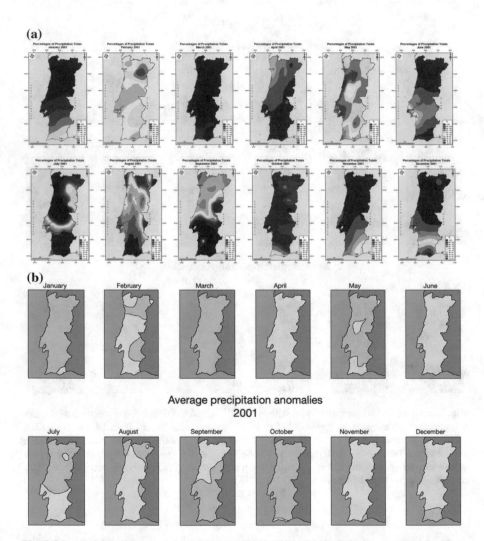

Fig. 3.26 **a** Precipitation anomalies in the year 2001 (IPMA 2018). **b** Representation of regions with less than normal rainfall from 1971 to 2000. Beige represents the region with the lowest rainfall and lilac represents the region where the rainfall was equal to or greater than the corresponding period from 1971 to 2000. *Source* Own elaboration and IPMA (2018)

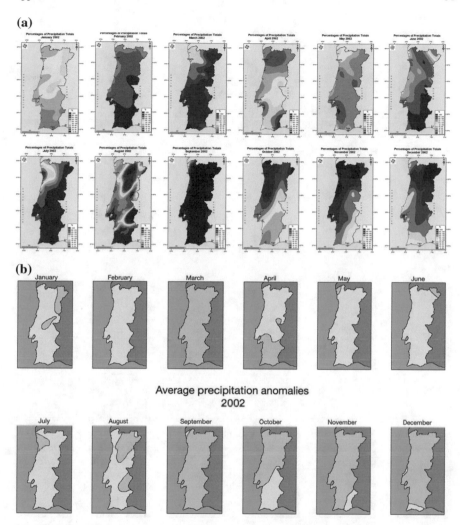

Fig. 3.27 a Precipitation anomalies in the year 2002 (IPMA 2018). **b** Representation of regions with less than normal rainfall from 1971 to 2000. Beige represents the region with the lowest rainfall and lilac represents the region where the rainfall was equal to or greater than the corresponding period from 1971 to 2000. *Source* Own elaboration and IPMA (2018)

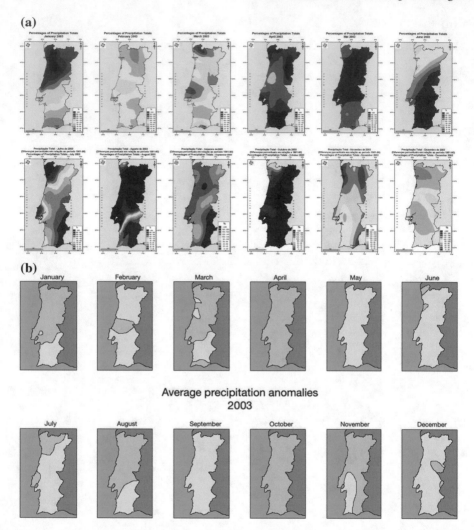

Fig. 3.28 a Precipitation anomalies in the year 2003 (IPMA 2018). **b** Representation of regions with less than normal rainfall from 1971 to 2000. Beige represents the region with the lowest rainfall and lilac represents the region where the rainfall was equal to or greater than the corresponding period from 1971 to 2000. *Source* Own elaboration and IPMA (2018)

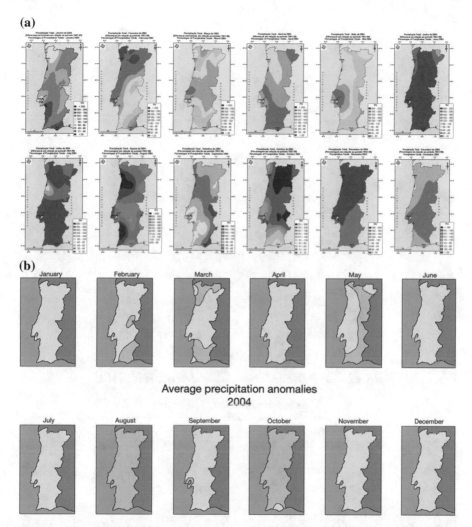

Fig. 3.29 a Precipitation anomalies in the year 2004 (IPMA 2018). **b** Representation of regions with less than normal rainfall from 1971 to 2000. Beige represents the region with the lowest rainfall and lilac represents the region where the rainfall was equal to or greater than the corresponding period from 1971 to 2000. *Source* Own elaboration and IPMA (2018)

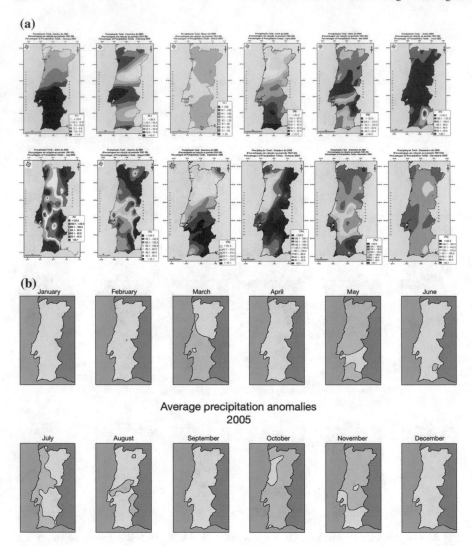

Fig. 3.30 a Precipitation anomalies in the year 2005 (IPMA 2018). **b** Representation of regions with less than normal rainfall from 1971 to 2000. Beige represents the region with the lowest rainfall and lilac represents the region where the rainfall was equal to or greater than the corresponding period from 1971 to 2000. *Source* Own elaboration and IPMA (2018)

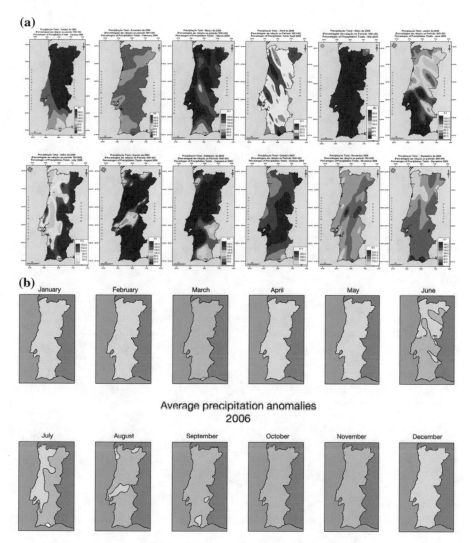

Fig. 3.31 a Precipitation anomalies in the year 2006 (IPMA 2018). **b** Representation of regions with less than normal rainfall from 1971 to 2000. Beige represents the region with the lowest rainfall and lilac represents the region where the rainfall was equal to or greater than the corresponding period from 1971 to 2000. *Source* Own elaboration and IPMA (2018)

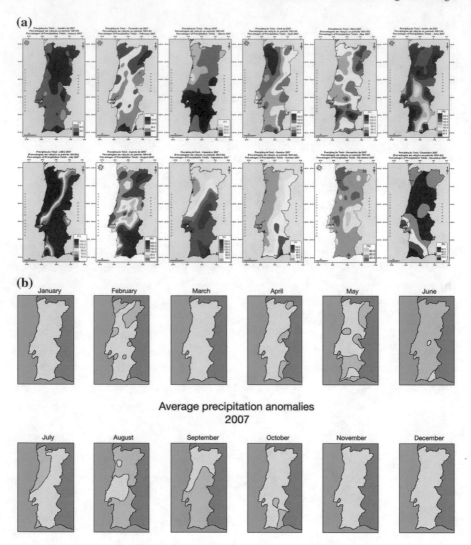

Fig. 3.32 a Precipitation anomalies in the year 2007 (IPMA 2018). **b** Representation of regions with less than normal rainfall from 1971 to 2000. Beige represents the region with the lowest rainfall and lilac represents the region where the rainfall was equal to or greater than the corresponding period from 1971 to 2000. *Source* Own elaboration and IPMA (2018)

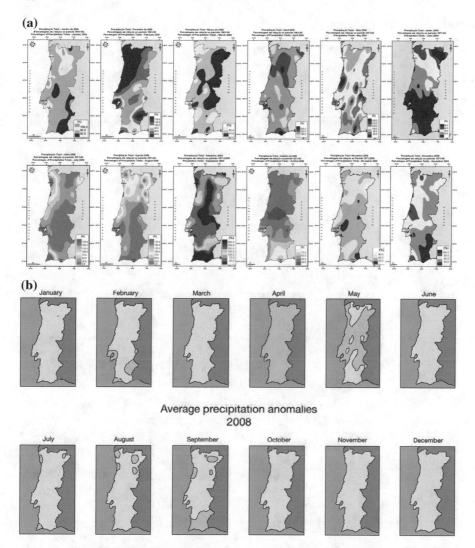

Fig. 3.33 **a** Precipitation anomalies in the year 2008 (IPMA 2018). **b** Representation of regions with less than normal rainfall from 1971 to 2000. Beige represents the region with the lowest rainfall and lilac represents the region where the rainfall was equal to or greater than the corresponding period from 1971 to 2000. *Source* Own elaboration and IPMA (2018)

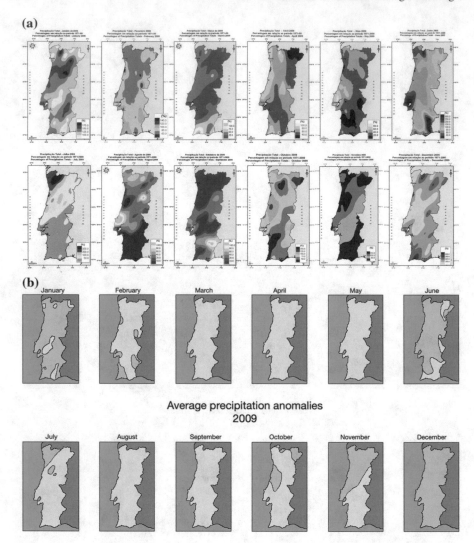

Fig. 3.34 a Precipitation anomalies in the year 2009 (IPMA 2018). **b** Representation of regions with less than normal rainfall from 1971 to 2000. Beige represents the region with the lowest rainfall and lilac represents the region where the rainfall was equal to or greater than the corresponding period from 1971 to 2000. *Source* Own elaboration and IPMA (2018)

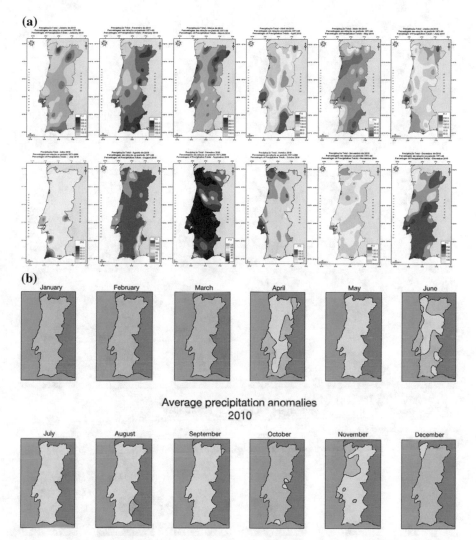

Fig. 3.35 **a** Precipitation anomalies in the year 2010 (IPMA 2018). **b** Representation of regions with less than normal rainfall from 1971 to 2000. Beige represents the region with the lowest rainfall and lilac represents the region where the rainfall was equal to or greater than the corresponding period from 1971 to 2000. *Source* Own elaboration and IPMA (2018)

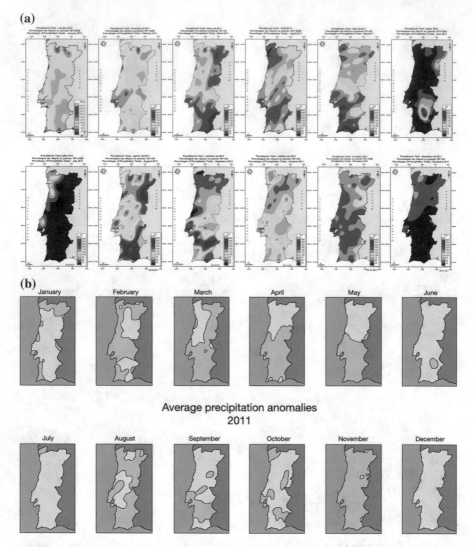

Fig. 3.36 **a** Precipitation anomalies in the year 2011 (IPMA 2018). **b** Representation of regions with less than normal rainfall from 1971 to 2000. Beige represents the region with the lowest rainfall and lilac represents the region where the rainfall was equal to or greater than the corresponding period from 1971 to 2000. *Source* Own elaboration and IPMA (2018)

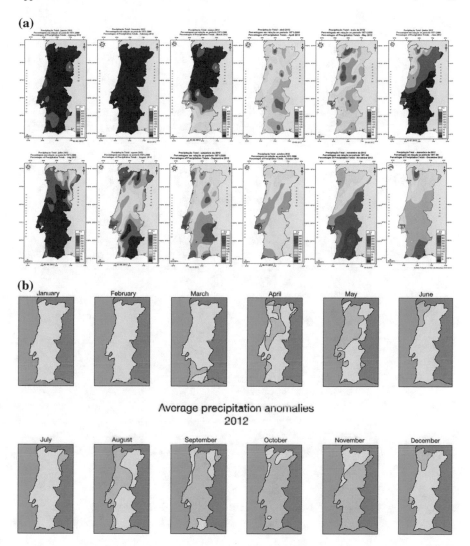

Fig. 3.37 a Precipitation anomalies in the year 2012 (IPMA 2018). **b** Representation of regions with less than normal rainfall from 1971 to 2000. Beige represents the region with the lowest rainfall and lilac represents the region where the rainfall was equal to or greater than the corresponding period from 1971 to 2000. *Source* Own elaboration and IPMA (2018)

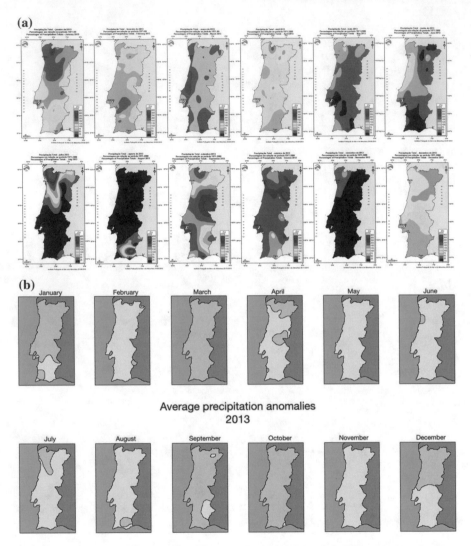

Fig. 3.38 a Precipitation anomalies in the year 2013 (IPMA 2018). **b** Representation of regions with less than normal rainfall from 1971 to 2000. Beige represents the region with the lowest rainfall and lilac represents the region where the rainfall was equal to or greater than the corresponding period from 1971 to 2000. *Source* Own elaboration and IPMA (2018)

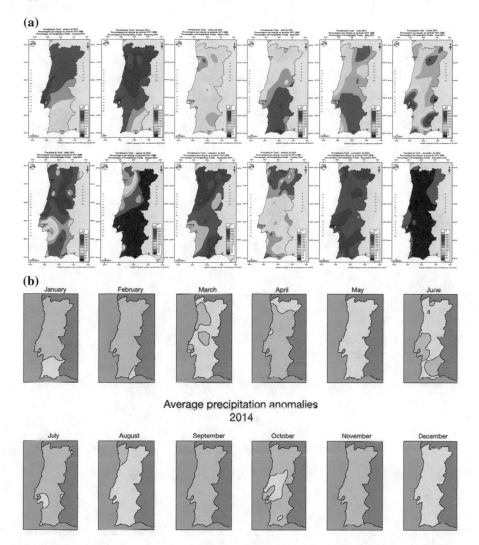

Fig. 3.39 a Precipitation anomalies in the year 2014 (IPMA 2018). **b** Representation of regions with less than normal rainfall from 1971 to 2000. Beige represents the region with the lowest rainfall and lilac represents the region where the rainfall was equal to or greater than the corresponding period from 1971 to 2000. *Source* Own elaboration and IPMA (2018)

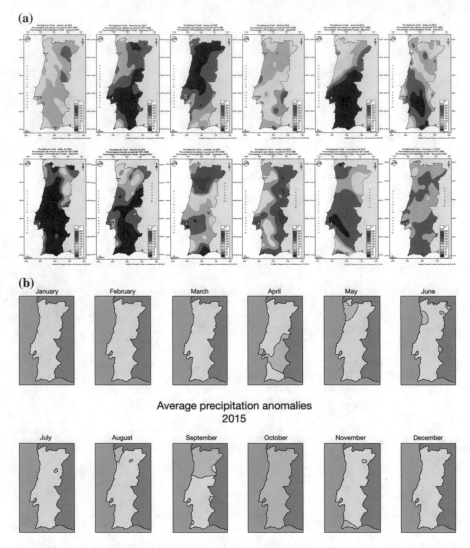

Fig. 3.40   a Precipitation anomalies in the year 2015 (IPMA 2018). **b** Representation of regions with less than normal rainfall from 1971 to 2000. Beige represents the region with the lowest rainfall and lilac represents the region where the rainfall was equal to or greater than the corresponding period from 1971 to 2000. *Source* Own elaboration and IPMA (2018)

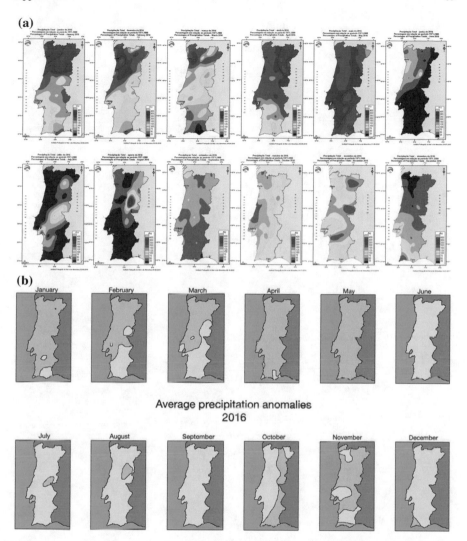

Fig. 3.41 a Precipitation anomalies in the year 2016 (IPMA 2018). **b** Representation of regions with less than normal rainfall from 1971 to 2000. Beige represents the region with the lowest rainfall and lilac represents the region where the rainfall was equal to or greater than the corresponding period from 1971 to 2000. *Source* Own elaboration and IPMA (2018)

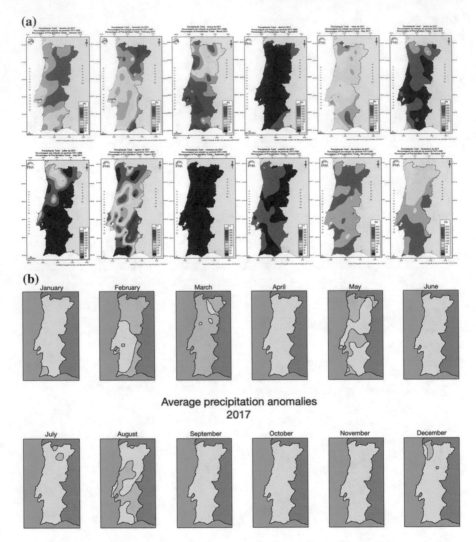

Fig. 3.42 a Precipitation anomalies in the year 2017 (IPMA 2018). **b** Representation of regions with less than normal rainfall from 1971 to 2000. Beige represents the region with the lowest rainfall and lilac represents the region where the rainfall was equal to or greater than the corresponding period from 1971 to 2000. *Source* Own elaboration and IPMA (2018)

References

Crowder SV (1987) A simple method for studying run–length distributions of exponentially weighted moving average charts. Technometrics 29(4):401–407

Giorgi F, Lionello P (2008) Climate change projections for the Mediterranean region. Glob Planet Chang 63(2–3):90–104

IPMA (s.d., 03/09/2018) O Clima em Portugal Continental. Available https://www.ipma.pt/pt/educativa/tempo.clima/index.jsp?page=clima.pt.xml

Pausas JG (2004) Changes in fire and climate in the eastern Iberian Peninsula (Mediterranean basin). Clim Change 63(3):337–350

Peel MC, Finlayson BL, McMahon TA (2007) Updated world map of the Köppen-Geiger climate classification. Hydrol Earth Syst Sci Dis 4(2):439–473

Pereira JS et al (2006) "Florestas e biodiversidade," Alterações climáticas em Portugal—cenários, impactos e medidas de adaptação (Projecto SIAM II). Gradiva, Lisbon, pp 301–343

TSF/LUSA (2018) Portugal é o 2.º melhor país europeu na luta contra alterações climáticas. TSF, ed. Portugal

Chapter 4
Impact of Climate Changes in Forest Development

Abstract By verifying the existence of a tendency for the occurrence of climatic anomalies, such as those previously verified, can lead to the occurrence of profound changes in the regulatory parameters of ecosystems, and thus promote the occurrence of changes in the forest space. These changes are addressed in this chapter, highlighting the emergence of diseases and pests, forest invasive species and also the increased risk of forest fires.

Keywords Forest plagues and diseases · Forest invasive species · Forest fires · Forest evolution

4.1 Framework

In Portugal, forests occupy approximately 3.2 million hectares, about 35% of the total area of the country (ICNF 2013). Portuguese forests provide many benefits and services for society, including clean water and air, recreational and recreational spaces, wildlife habitats, carbon sequestration and storage, climate regulation, and a variety of forest products with major impacts on the economy (Mendes 1997; Ramos 2012).

Climate influences the structure and function of forest ecosystems and plays a key role in forest health. A changing climate can intensify many of the threats to forests, such as outbreaks of pests, fires, drought and the very development of populations there (Sarmento and Dores 2013).

Climate change directly and indirectly affects the growth and productivity of forests through changes in temperature, precipitation, climate and other factors. In addition, levels of high levels of carbon dioxide can also affect plant growth. These changes influence the complex forest ecosystems in various ways (Bongers et al. 2015).

Together with the impacts resulting from the effects of climate change, forests face impacts due to the development of land management, namely due to their use and occupation, periodic forest fires and atmospheric pollution. Although it is difficult to separate the effects of these different factors, the combined impact is already

© The Author(s), under exclusive license to Springer Nature Switzerland AG 2020
L. J. R. Nunes et al., *Climate Change Impact on Environmental Variability in the Forest*, SpringerBriefs in Environmental Science, https://doi.org/10.1007/978-3-030-34417-7_4

causing changes and changes in Portuguese forests. As these changes are expected to continue in the coming decades, some of the economic aspects provided by forests may be compromised in the short term (Pereira 2016).

4.2   Forest Growth and Productivity

As mentioned earlier, there are several aspects related to climate change that are very likely to significantly affect the growth and productivity of forest species. Examples of such factors are the rise in temperature, changes in precipitation levels and increases in the concentration of carbon dioxide (Lempereur et al. 2015).

The rise in temperature generally increases the duration of the plant growing season. This factor also contributes to the change in the geographic dispersion of some tree species. In this way, the habitats of some types of trees tend to move to the north or to higher altitudes. Other species will be at risk locally or regionally if the conditions in their current geographic ranges are not the most appropriate. For example, species that currently exist only on top of mountains in some regions may de-saparecer as the climate warms up, since these species can not evolve to a higher altitude (Clark et al. 2016).

Climate change is most likely to increase the risk of drought in some areas and the risk of extreme rainfall and flooding in other regions (Matos 2017). Increased temperatures change the defrost time, affecting the seasonal availability of water. Although many trees achieve up to a certain degree of drought, rising temperatures can make future droughts more damaging than they have in the past. In addition, drought increases the risk of forest fires, since dry trees and shrubs provide fuel for the fires. Drought also reduces sap trees' ability to produce sap, which protects them from destructive insects and disease (Bennett et al. 2015).

Carbon dioxide is needed for photosynthesis, the process by which green plants use sunlight to grow. With sufficient water and nutrients, increases in atmospheric CO_2 concentration may allow trees to have higher growth rates, which may alter the distribution of tree species. Growth will be higher on nutrient rich soils, with no water limitation, and will decrease with reduced fertility and water supply (Doughty et al. 2015).

4.3   Plagues, Invasive Species and Forest Fires

Climate change can change the frequency and intensity of negative impacts on the forest, such as pest outbreaks, invasive species proliferation, forest fires and storm surges. These disturbances can reduce the productivity of the forest and alter the distribution of the flower-like species. In some cases, forests may recover from a disturbance, but in other cases, existing species may evolve or disappear. In these cases, the new plant species that colonize a certain area, create a new type of forest (Anderegg et al. 2015).

According to the opinion of several authors, with temperatures rising as the climate changes and warms, the insects will become more active, and consequently more ravenous and abundant, increasing the probability of occurrence of insect pests. In this way damage to crops, both agricultural and forestry, is bound to increase (Deutsch et al. 2018).

This argument gains strength, as the increase in temperature accelerates the metabolism and reproduction of insects. The same authors estimate that each degree of increase in temperatures will mean increased crop damage from 10 to 25% (Deutsch et al. 2018).

Unlike mammals and birds, insects heat or cool according to their environment. When an insect warms up, its metabolism accelerates. The faster you burn energy, the more voraciously the insect feeds and the sooner you can reproduce. The analysis of this information allowed to conclude that the population growth rates are not very different among the different types of insects, enabling the development of a mathematical model that simulates the growth of insect populations. This simulation, later, allows to infer the damage caused by these same populations of insects in the crops, both agricultural and forestry (Deutsch et al. 2018).

Tropical insects are usually close to the upper limit of their tolerance to temperature, so it can be concluded that it is not in these regions that the greatest variations in insect populations will be observed. In areas with a more temperate climate, insects can significantly accelerate their activity, causing more damage to crops (Fig. 4.1) (Deutsch et al. 2018).

Variations in temperature may encourage or discourage insect species from invading new territories. Temperatures can also affect the parasites that attack these same insects that attack crops, so the end result will greatly depend on the ability of all stakeholders to evolve and adapt to the new reality (Deutsch et al. 2018).

A number of biotic and abiotic agents have been identified in Portugal that are capable of causing physiological imbalances responsible for changes in the development of trees and that may be associated with the high frequency with which

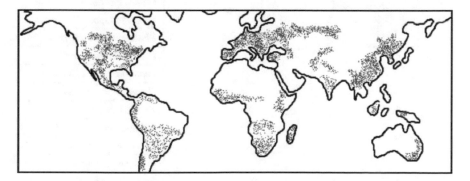

Fig. 4.1 Prediction of regions where there will potentially be a greater percentage increase in crop yield losses. Adapted from Milius (2018)

forest fires occur, which could lead to the appearance of pests and diseases. Table 4.1 presents some of these pests that constitute major sanitary problems for the forest.

The xylophagous or subcortical insects that attack the trunk, can be considered as the most serious pests, since they block the circulation of sap, putting at risk the survival of the tree. Some of these insects also have the ability to inoculate agents that contribute to the weakening and death of trees (Naves 2016).

Insect attacks on leaves/needles do not normally jeopardize tree survival (with the exception of very severe attacks on young stands), as shown in Fig. 4.2, because of its severity, since the young tree was totally stripped of needles, driving the tree to death. In other situations, for example with older and larger trees, canopy regeneration may occur, although it may show a decrease in its annual growth rate, since the energy reserves will be channeled for the renewal of the foliage (Naves 2016).

Table 4.1 Main pests of the forest in Portugal

Forest species	Scientific name of the plague
Eucalyptus	*Gonipterus platensis*
	Phoracantha semipunctata
Cork oak and holm oak	*Lymantria dispar*
	Periclista spp.
	Tortrix viridana
	Curculio elephas
	Cydia splendana
	Coroebus undatus
	Coroebus florentinus
	Platypus cylindrus
	Xyleborus spp.
Maritime pine and stone pine	*Thaumetopoea pityocampa*
	Pineus pini
	Cinara maritima
	Leucaspis spp.
	Pissodes validirostris
	Dioryctria mendacella
	Leptoglossus occidentalis
	Orthotomicus erosus
	Tomicus spp.
	Ips sexdentatus
	Dioryctria sylvestrella
	Pissodes castaneus
	Monochamus galloprovincialis

Adapted from Naves (2016)

Fig. 4.2 Young pine tree (Pinus pinaster) after the attack of an insect, the Pissodes castaneus. It is a species of insect Coleoptera polyphagus belonging to the family Curculionidae

This situation is more critical in the resinous species since its leaf surface is renewed in a slower way than in the hardwood species. However, it is the pests and diseases introduced in Portugal that cause the most damage to the forest. An example of this is the introduction of the wild pine wood nematode (Bursaphelenchus xylophilus) in 1999, which is currently the most serious problem of forest health in Portugal, and is responsible for enormous economic losses, not only due to the dead pine trees, but also due to the strong import restrictions on pine wood (Naves 2016).

Eucalyptus also suffered from the introduction of exotic species, such as mortality caused by the eucalyptus borer (Phoracantha semiounctata) in the late 1990s, mainly in Beiras and Alentejo, and, more recently, the damage caused by the defoliant weevil in the North and Center regions (Naves 2016).

Recently, other exotic insects have been registered in Portugal, such as the eurasian golden-bug (Thaumastocoris peregrinus) and the chestnut-horned wasp (Dryocosmus kuriphilus) (Ribeiro 2017). Its medium and long-term impact in the national territory is not yet known, but because of the risk of being an emerging pest, action plans have already been implemented to control it (Naves 2016).

The lack of natural controls such as predators or pathogens, as well as the fact that tree defenses are unsuitable for these harmful agents, may allow the insects to spread. Climate change could contribute to an increase in the severity of future insect outbreaks. Rising temperatures may allow some insect species to develop more rapidly, alter their seasonal life cycles, and expand their activity to other latitudes than usual (Endara-Agramont et al. 2013; Arnaldo et al. 2011).

An important issue is that species of plants considered invasive can occupy the space of native vegetation, since in principle these species, which are outside their original habitat, do not have natural predators in the new environment (Arnaldo et al. 2011; Mortensen et al. 2009).

Climate change can benefit invasive plants that are more tolerant of new environmental conditions than native plants (Figs. 4.3 and 4.4).

In 2017, forest fires consumed more than 442,000 ha of forest in Portugal, causing 104 fatalities and more than 600 million euros in losses (Fernandes et al. 2017; Silva 2017). High temperatures and drought conditions during the early summer contributed to this tragic scenario (Bugalho and Pessanha 2018).

The forest fires of 2017 were the largest catastrophe in the country in terms of the number of fatalities since the fateful floods of 1967 (Espada 2017). Forest fires cause a feeling of powerlessness to the people, who consider them almost a fatality intrinsic to the Portuguese forest and forest, to the point that the term "fire season" is generalized, as if it were naturally part of the calendar as another season (Martins et al. 2017). Since 2000, 200 people have died as a result of forest fires in Portugal (LUSA 2017). This year, in addition to the fatalities, about 300 people were injured and entire communities still suffer post-traumatic stress (Carvalho et al. 2018). In that fateful year, the situation reached a level never before seen because, in addition to the victims, the territory suffered damages almost irreparable (Amaro and Barroco 2018).

As mentioned earlier, hundreds of thousands of hectares of forest have disappeared, accounting for more than half of the area burned in Europe in 2017 (E Commission 2018), and a great historical landmark, since the fire decimated a significant part of the Pinhal do Rei in Leiria, from where the legend must have left the mythical wood that served to build the ships and the caravels that took the Portuguese in the period of maritime expansion (Martins et al. 2017). But it was not only the past that was destroyed, as the industrial and agricultural sectors were also largely hit, putting a high number of jobs at risk.

However, this event is not new and in 2017 it only reached the contours that it reached due to the absurd number of deaths and injuries, since in previous years similar phenomena were observed, except for the number of victims to be regretted (Fig. 4.5). The effect of climate change is expected to contribute to increasing the extent, intensity and frequency of forest fires in certain areas of the country. Warmer spring and summer temperatures, coupled with reduced availability of water, dry woody materials in forests and increase the risk of forest fires.

Fires can also contribute to the very phenomenon of climate change, as they can cause large and rapid releases of carbon dioxide into the atmosphere (Phillips 2018).

It is important to note that fire activity is not only determined by drought as a structural basis and by meteorology as the conjunctural basis of risk. The continued aridity has effects of greater temporal availability of all the vegetation to burn, resulting in campaigns of continuous or extended fires, creating the conditions for large fires more easily and quickly than under the "normal" regime of situations (Independente 2017).

Fig. 4.3 Mimosas stains (Acacia dealbata) on a slope of Serra da Estrela. This photograph, taken in the town of Casal do Rei, in the União de Freguesias de Cabeça and Vide, in the municipality of Seia, shows how this invasive species is gaining ground, especially in relation to the dominant species in the area. (Pinus pinaster), but also to the other species existing on the site. It is a recovering forest area after a major forest fire occurred in 2005

Fig. 4.4 Occurrences of Hakea spiny (Hakea sericea), a thorny shrub of Australian origin, which can reach 4–5 m in height and has a compact appearance, so it was often used to form protective hedges (Torres Ribeiro and Freitas 2010). It was also most likely the initial mode of propagation. This photograph, taken in the town of Casal do Rei, in the União de Freguesias de Cabeça and Vide, in the municipality of Seia, shows how this invasive species proliferates along the roads and roads, taking advantage of the fact that they are sites more exposed to Sun light

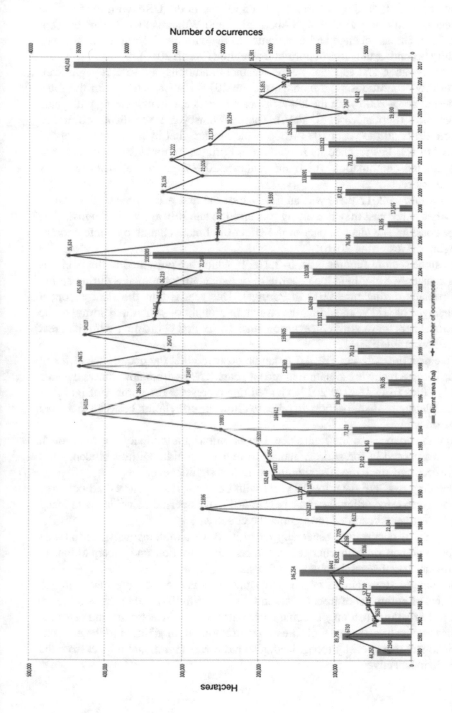

Fig. 4.5 Area burned in hectares and number of ignitions occurred in Portugal between 1980 and 2017. Fonte: elaboração própria com os dados obtidos em (ICNF 2013)

The concept of "aridity" of fuel has been related in the USA with the increase of forest fires in size and severity (Abatzoglou and Williams 2016). This concept is related to climate change and contributes to the extreme fire regime of the past gradually becoming the current normal fire regime (Werth 2016).

Spring weather, and especially June weather conditions, led to very significant fuel "aridity" conditions. The months of June 2015–2017 together with the four-year period 2003–2006 were the hottest since record keeping, coinciding with years of larger fires (Fernandes et al. 2016). There is, however, a significant difference between 2017 and the years 2003–2006, which means that in 2017 the concept of "aridity" of fuel has reached its maximum in Portugal by making a warm June to hot spring, this combination which differentiates 2017 from the years of great fires of the recent past.

As a result, in 2017 there was an advance of the first major fires of the year to June, when in the past they had always occurred in late July and early August. This change constitutes a clear impact of the effect of climate change on contemporary fire regimes (Westerling 2016).

As can be seen in Figures 3.9a, b–3.22a, b, which are presented in the Annexes to this chapter, we highlight the occurrence of temperature anomalies during spring and especially during June 2017 in Portugal. This year is thus the clear exponent of a new type of arid year, adding the structural drought of winter and spring to the exceptional heat wave with epicenter on June 16–19 and coinciding with the great fires that occurred in 2017.

In the data presented in Fig. 4.5 it can be observed that the occurrence of forest fires has been a recurring situation in recent years. The figure below shows the area burned in fires from 1980 to 2017. The red line represents the number of ignitions occurred during the same period, in order to allow the correlation between them and the burned area.

From the analysis of the figure can also be inferred a tendency for the growth of the area burned in forest fires, mainly from the 90s, although this tendency has already occurred since the 80s, although with less significance.

According to ICNF data, the 10 years with the highest area burned all occurred after this period, including the peak year in 2017. This period coincides with many of the hottest years recorded throughout the country.

This type of occurrence, especially when it occurs simultaneously, can interact with one another, or with changes in temperature variation and precipitation, to increase the risk of several impacts on forests.

For example, drought can weaken trees and make the forest more susceptible to forest fires or outbreaks of insect pests and diseases. Similarly, after the occurrence of a forest fire, the forest may become more vulnerable to insect pests and diseases.

This permanent occurrence of forest fires in Portugal, in addition to the environmental impact that is easily recognized, also has a very significant impact from the economic perspective.

In fact, as already mentioned, the trivialization of the "fire season" as if it were a seasonally repeatable period of the Portuguese year, led to the inclusion of significant amounts of costs in the state budget for fire prevention forest, in addition to the creation of a device more and more modern and prepared to fight the said fires.

Table 4.2 presents the costs related to the occurrence of forest fires from 1980 to 2017. From the simple analysis of the data, there is a progressive increase in costs associated with the growth of the burned areas.

The amount presented represents the sum corresponding to the amount spent each year on the prevention, combat and costs associated with the damages caused by forest fires.

This phenomenon of forest fires in Portugal, which is now beginning to be understood that can be boosted by climate change, does not however originate in these changes or even natural causes in the overwhelming majority of occurrences.

Nowadays, the cause of the origin of the great majority of the forest fires is already investigated, being therefore possible to refute the thesis of the natural origin for these occurrences.

Both the ICNF and the Nature and Environment Protection Service (SEPNA), belonging to the National Republican Guard (GNR), have the human and technical capacity to analyze the causes that caused a certain ignition and disseminate the results of the investigations in documents that are available at their respective websites, which can be followed at the addresses, respectively www.icnf.pt and http://www.gnr.pt/atrib_SPENA.aspx.

From the analysis of these data we obtain the results that are presented in Table 4.3.

From the analysis of the data presented in Table 4.3, it can be observed that in the vast majority of cases, forest fires have an accidental, negligent or malicious origin, originated in human activity, and the role of the natural phenomena to which responsibilities, are clearly very insignificant.

This group includes, for example, the occurrence of dry thunderstorms, very frequent in the summer. However, as can be seen in the previous table, with a very insignificant weight in the overall number of occurrences.

Table 4.2 Annual costs associated with forest fires

Year	Occurrences (nr.)	Forest area (m²)	Total costs (€)
1980	2.349	44.251	€96.024.670
1981	6.730	89.798	€194.861.660
1982	3.626	39.556	€85.836.520
1983	4.542	47.811	€103.749.870
1984	7.356	52.710	€114.380.700
1985	8.441	146.254	€317.371.180
1986	5.036	89.522	€194.262.740
1987	7.705	76.268	€165.501.560

(continued)

Table 4.2 (continued)

Year	Occurrences (nr.)	Forest area (m^2)	Total costs (€)
1988	6.131	22.434	€48.681.780
1989	21.896	126.237	€273.934.290
1990	10.745	137.252	€297.836.840
1991	14.327	182.486	€395.994.620
1992	14.954	57.012	€123.716.040
1993	16.101	49.963	€108.419.710
1994	19.983	77.323	€167.790.910
1995	34.116	169.612	€368.058.040
1996	28.626	88.857	€192.819.690
1997	23.497	30.535	€66.260.950
1998	34.675	158.369	€343.660.730
1999	25.473	70.613	€153.230.210
2000	34.107	159.605	€433.000.000
2001	26.947	112.312	€319.000.000
2002	26.576	124.619	€394.000.000
2003	26.219	425.839	€1.303.000.000
2004	22.165	130.108	€402.000.000
2005	35.824	339.089	€756.746.827
2006	20.444	76.058	€132.001.898
2007	20.316	32.595	€37.109.004
2008	14.930	17.565	€22.371.685
2009	26.136	87.421	€86.259.214
2010	22.026	133.091	€183.911.947
2011	25.222	73.829	€80.557.921
2012	21.179	110.232	€196.227.660
2013	19.294	152.690	€208.337.840
2014	7.067	19.930	€27.503.169
2015	15.851	64.412	€119.406.200
2016	13.079	160.490	€460.000.000
2017	16.981	442.418	€616.000.000
Total		4419.166	€9,589,826.075

Source Own elaboration based on data available in ICNF (2013),
Toimil et al. (2018)

Table 4.3 Type of causes of forest fires investigated by GNR/SEPNA in 2016

Causes	Percentage of total occurrences (%)
Undetermined	34.7
Use of fire	25.2
Arsonism	21.6
Rekindles	13.6
Accidentals	3.5
Naturals	0.7
Struturals	0.6

Adapted from ICNF (2013)

Fig. 4.6 Panoramic photograph taken on 05/12/2017 from ESA—Ponte de Lima Higher School of Agriculture of the Polytechnic Institute of Viana do Castelo, where it is possible to observe 9 fires in rural environment occurring simultaneously. *Source* Elaboration own

Of all the frequent causes, it continues to be the "use of fire", mainly in the form of burnings for the elimination of agricultural and forest residues, the main cause for the occurrence of forest fires (Fig. 4.6).

References

Abatzoglou JT, Williams AP (2016) Impact of anthropogenic climate change on wildfire across western US forests. Proc Natl Acad Sci 113(42):11770–11775

Amaro S, Barroco C (2018) O Impacto dos Incêndios no Turismo em Espaço Rural na Região Centro. Turismo no Centro de Portugal-potencialidades e tendências, pp 155–175

Anderegg WR et al (2015) Pervasive drought legacies in forest ecosystems and their implications for carbon cycle models. Science 349(6247):528–532

Arnaldo PS, Oliveira I, Santos J, Leite S (2011) Climate change and forest plagues: the case of the pine. Forest Syst 20(3):508–515

Bennett AC, McDowell NG, Allen CD, Anderson-Teixeira KJ (2015) Larger trees suffer most during drought in forests worldwide. Nature Plants 1(10):15139

Bongers F, Chazdon R, Poorter L, Peña-Claros M (2015) The potential of secondary forests. Science 348(6235):642–643

Bugalho L, Pessanha L (2018) Relação da área ardida em Portugal com o risco de fogo florestal ICRI. Acta de las Jornadas Científicas de la Asociación Meteorológica Española (32)

Carvalho AS, Marques S, Rosário F (2018) Gone with the fire: how family physicians in central Portugal experienced the aftermath of the Great Fire of October 15, 2017. Acta Med Port 31(1):7–8

Clark JS et al (2016) The impacts of increasing drought on forest dynamics, structure, and biodiversity in the United States. Glob Change Biol 22(7):2329–2352

Deutsch CA et al (2018) Increase in crop losses to insect pests in a warming climate. Science 361(6405):916–919

Doughty CE et al (2015) Drought impact on forest carbon dynamics and fluxes in Amazonia. Nature 519(7541):78

E Commission (2018) European Forest Fire Information System (EFFIS). Available: http://effis.jrc.ec.europa.eu/

Endara-Agramont AR, Calderón-Contreras R, Nava-Bernal G, Franco-Maass S (2013) Analysis of fragmentation processes in high-mountain forests of the centre of Mexico. Am J Plant Sci 4(03):697

Espada MH (2017) Não falar do mau cheiro dos cadáveres: como Salazar escondeu 700 mortos. Available: https://www.sabado.pt/portugal/detalhe/cheias-de-1967-a-tragedia-esquecida-que-matou-centenas

Fernandes PM, Monteiro-Henriques T, Guiomar N, Loureiro C, Barros AM (2016) Bottom-up variables govern large-fire size in Portugal. Ecosystems 19(8):1362–1375

Fernandes S, Meira Castro A, Lourenço L (2017) Duas décadas de investigação das causas de incêndios florestais em Portugal continental. In IV Congresso Internacional de Riscos "Riscos e Educação"

ICNF (2013) IFN6-Áreas dos usos do solo e das espécies florestais de Portugal continental. Resultados preliminares. Instituto da Conservação da Natureza e das Florestas Lisboa (ed)

Independente CT (2017) Análise e apuramento dos factos relativos aos incêndios que ocorreram em Pedrogão Grande, Castanheira de Pera, Ansião, Alvaiázere, Figueiró dos Vinhos, Arganil, Góis, Penela, pampilhosa da Serra, Oleiros e Sertã, entre 17 e 24 de junho de 2017. Assembleia da República, Lisboa2017, Available: https://www.parlamento.pt/Documents/2017/Outubro/RelatórioCTI_VF%20.pdf

Lempereur M et al (2015) Growth duration is a better predictor of stem increment than carbon supply in a Mediterranean oak forest: implications for assessing forest productivity under climate change. New Phytol 207(3):579–590

LUSA (2017) Cronologia: pelo menos 200 mortos deste 2000 em incêndios, 2017 é o pior ano. Available: https://www.dn.pt/lusa/interior/cronologia-pelo-menos-200-mortos-desde-2000-em-incendios-2017-e-o-pior-ano-8848644.html

Martins PSGC, Beleza J, Duarte JS (2017) Um fogo que não se apaga. In Expresso (ed)

Matos ATLS (2017) Gestão de espaços florestais em áreas protegidas de montanha: o caso do Perímetro Florestal de Manteigas

Mendes AC (1997) Estimativa do valor económico da floresta portuguesa. In Congresos Forestales

Milius S (2018) As temperatures rise, so do insects' appetites for corn, rice and wheat. Available: https://www.sciencenews.org/article/temperatures-insects-appetites-crop-damage-global-warming

Mortensen DA, Rauschert ES, Nord AN, Jones BP (2009) Forest roads facilitate the spread of invasive plants. Invasive Plant Sci Manage 2(3):191–199

Naves ESLBP (2016) Principais pragas florestais em Portugal: Como atuar? Mundo Rural, pp 36–37

Pereira JS (2016) O futuro da floresta em Portugal. Fundação Francisco Manuel dos Santos

Phillips R et al (2018) Indiana's future forests: a report from the Indiana climate change impacts assessment

Ramos ED (2012) Certificação Florestal. Custos e Benefícios da Certificação da Gestão Florestal em Portugal

Ribeiro M (2017) Sustentabilidade dos recursos florestais em Portugal: o que aprendemos com o pinheiro-bravo. In 8° Congresso Florestal Nacional. Raízes do Futuro,Floresta em Português

Sarmento EM, Dores V (2013) Desafios para a Gestão: Ambiental da Fileira Florestal em Portugal. Silva Lusitana 21(1):1–19

Silva CC (2017) 2017 foi o ano em que mais ardeu nos últimos 10 anos—4 vezes mais do que o habitual. Available: https://www.publico.pt/2017/11/10/sociedade/noticia/2017-foi-o-ano-em-que-mais-ardeu-nos-ultimos-dez-anos–quatro-vezes-mais-que-o-habitual-1792180

Toimil A, Díaz-Simal P, Losada IJ, Camus P (2018) Estimating the risk of loss of beach recreation value under climate change. Tour Manag 68:387–400

Torres Ribeiro K, Freitas L (2010) Impactos potenciais das alterações no Código Florestal sobre a vegetação de campos rupestres e campos de altitude. Biota Neotropica 10(4):239–246

Werth PA et al (2016) Synthesis of knowledge of extreme fire behavior: volume 2 for fire behavior specialists, researchers, and meteorologists. Gen Tech Rep PNW-GTR-891. Portland, OR: US Department of Agriculture, Forest Service, Pacific Northwest Research Station, vol 891, 258 p

Westerling AL (2016) Increasing western US forest wildfire activity: sensitivity to changes in the timing of spring. Phil. Trans. R. Soc. B 371(1696):20150178

Chapter 5
Discussion and Conclusions

Abstract With the foreseeable occurrence of pests and diseases in the forest, with the exponential growth of forest invasive species, and with the increased risk of forest fires, it is necessary to make an assessment and a state of play on the current state of the forest in Portugal. In this way, it is intended to find potential solutions to prevent the problems that will surely happen. In this chapter this analysis and discussion is made, but recommendations are also made, for example, for the prevention of forest fires through the adoption of measures such as the extension of the surveillance period beyond the traditional period, since with the occurrence of anomalies It is foreseeable that these may arise even outside the so-called "normal" season for forest fires.

Keywords Mitigation measures · Forest fires prevention · Portuguese forest · Global changes

The planet's climate has been undergoing major changes for several decades. The IPCC report indicates that climate warming is evident and that most is probably due to the increase in GHG concentrations caused by human activities, such as widespread use of fuels, decomposition of municipal or animal waste and changes in the occupation of the soil.

There is already irrefutable proof of this change. The temperature of the atmosphere at the surface level has undergone a progressive warming from the beginning of the industrial age to the present day of approximately 0.6 °C on average, with an even greater increase in some areas such as the poles or the Mediterranean region.

The hottest years since recordings have occurred since 1990, as well as major seasonal changes, such as the decline of icy surfaces, rising sea levels, changes in the overall circulation flow of marine currents, and so on.

The frequency and severity of extreme weather events has increased. There is a more frequent occurrence of floods, heat and cold waves and periods of prolonged drought. An example of this is the news and constant warnings of hurricanes and storms of extreme force, which always carry high economic and personal damages.

Some high mountain species are already disappearing, unable to adapt to the rapid climatic changes of their habitats. Several coastal populations are threatened by rising sea levels as a result of the melting of large masses of ice and rising sea temperatures.

© The Author(s), under exclusive license to Springer Nature Switzerland AG 2020
L. J. R. Nunes et al., *Climate Change Impact on Environmental Variability in the Forest*, SpringerBriefs in Environmental Science,
https://doi.org/10.1007/978-3-030-34417-7_5

There is a full consensus on the part of the scientific community when it comes to attributing to the increased concentration of GHG generated by human activities the greater responsibility for the phenomenon of climate change. The reality is that without the natural presence of some of these gases in the atmosphere, such as water vapor and CO_2, creating the known greenhouse effect, Earth would be a very different place from what is known today, with average temperatures far below current.

Through the greenhouse effect, some gases retain the radiation emitted by the surface of the Earth, preventing them from getting lost into space. Without GHG, an average temperature on the Earth's surface is estimated to be around $-19\,°C$, instead of the current $+14\,°C$. The natural greenhouse effect makes life possible on the planet. However, the burning of fossil fuels, the destruction of forests, changes in land use, the production of waste and the emission of certain artificial gases are factors that reinforce the greenhouse effect, threatening the complex climate structure.

Since the beginning of the industrial revolution, when large quantities of fossil fuels began to be burned to meet the energy needs of industrial processes, so far the amount of CO_2 in the atmosphere has increased steadily. Likewise, other anthropogenic GHGs also increased their concentration in the atmosphere considerably.

Demographic growth and the current socio-economic model put great pressure on the self-regulating capacity of the atmosphere, which is leading to a situation close to its limits and, according to some scientists, to overcome them.

The main causes for GHG emissions vary according to the regions of the planet. Thus, in the northern hemisphere, the main causes are associated with energy production, industrial production and transportation, while in the southern hemisphere the main causes are associated with the change in land use, namely through the conversion of extensive forest areas into agricultural land or pasture.

It should be noted that in recent years there has been an effort in industrialized countries, with some success in some cases like Portugal, to reduce carbon dioxide emissions. The reasons for this reduction are the introduction of more efficient technologies, the use of renewable energies, the increase of the service sector and the shift of the most polluting companies to the least developed countries.

However, the steady growth of these industrialized economies, as well as the significant increase in emissions in other sectors, such as transport and the domestic sector, have made the total amount of GHG emissions of human origin increase considerably in recent years.

Forests, like other natural ecosystems, are as susceptible to climate change as other sectors (such as agriculture, for example, which are also highly vulnerable to climate change and the environment). Unlike other sectors, where financial resources and technology can directly contribute to increasing the adaptive capacity of affected systems, natural forests depend on their own natural ability to adapt.

Adding to all this human pressure, pressure for development, and climate change itself, it is highly likely that the ability of forestry systems to adapt to the new situation quickly and efficiently will be exceeded. It is expected that different forest systems will have different sensitivities to changes in climate. It is therefore important to take into account that the conservation of forests for other uses and their ability to sequester carbon dioxide can contribute significantly to carbon dioxide emissions in the future if forest systems are affected by natural or Men.

The way ecosystems respond to climate change is usually guided by two paradigms: evolution and adaptation, as seen in previous sections. In the first one it is assumed that there will be a migration of ecosystems to other regions, almost intact, just looking for new locations where climatic and environmental conditions reproduce those where they are currently. The second paradigm assumes that as climate and other factors change, ecosystems will change in the very place where they are today, which will interfere with both the variety of species and their position in the ecosystem.

In addition to the intrinsic value of natural ecosystems, ecosystems of all kinds, from the most natural to the most intensively managed, offer a variety of benefits to society at large. Some of the products originating from these ecosystems enter the market and contribute directly to economic development. For example, forests are a source of raw materials for a number of industries, such as the production of biomass pellets, pulp production, the production of wood pellets and the production of furniture. Forest ecosystems also provide a number of benefits to society, such as their role in regulating water flows, preventing erosion, maintaining biodiversity and temporary storage of carbon, which can be as long as the longevity of forest species and the more extensive the forestry operation recommended, for example, in the Portuguese case, using native species such as oaks, holm oaks or cork oaks.

Changes in soil cover caused by climate change can have a number of impacts on these benefits, such as the ability of these systems to stabilize the landscape against erosion or sequester carbon dioxide. Even in regions where the amount of existing vegetation is expected to increase as a result of higher precipitation rates and increased growth due to the higher concentration of atmospheric carbon dioxide may lead to an increase in the frequency and intensity of fires during a longer summer period. The increase in the occurrence of fires is already a threat not only to the vegetation cover, but also to the residential structures that are built in the rural areas, which are increasingly vulnerable. For this reason, it is very plausible that changes caused to natural ecosystems by climate change affect this set of benefits normally associated with forest activity.

As seen in the previous sections, the burning of fire, or simply the use of fire, is the main cause of occurrence of forest fires in Portugal, and its control and, if necessary, prohibition are urgently needed. Although it is an ancestral practice, with the worsening of climatic conditions, especially during the summer period, but which, as we have seen previously, now extendable also in the spring and autumn, the risk associated with this practice has grown exponentially. Most likely, the need to extend the so-called "forest fires season" will arise in the short term, where all means of combat are on alert so that they can react in a timely manner to any emergency.

As an example of this need, it is possible to present the outbreaks of fire already occurring during the month of October 2018, in which, following the anomalies of the average air temperature and precipitation in September and October 2018,

Fig. 5.1 Forest fire occurred on October 3, 2018, which demonstrates the need to rethink the need to extend the "forest fire season". *Source* Own elaboration

it is possible to present occurrences of forest fires with some intensity, such as the occurrence of October 3, 2018, in the municipality of Oliveira de Azeméis, in the parishes of Pinheiro da Bemposta, Travanca and Palmaz (Fig. 5.1).

Almost all climate models anticipate a decrease in the amount of precipitation in various parts of the globe, but they emphasize with particular emphasis the effects on the Mediterranean climate regions. The parallel effect of increasing temperature and reducing precipitation may lead to a significant decrease in the amount of soil moisture.

According to the information provided in the previous sections, there is already a tendency for climatic anomalies related to the increase of average air temperature in Continental Portugal, associated with the occurrence of anomalies related to the amount of precipitation in the several months of the year. That is, the trend indicates a strong probability of occurrence of periods of temperature increase associated with low levels of precipitation.

This lack can make trees more fragile from insect pest attacks and diseases, as well as increasing the likelihood of forest fires occurring. The frequency and intensity of these occurrences will determine the type and rate of land cover conversion to a new state, for example the replacement of a forest composed of native species, by another one composed of invasive species. In Portugal, infestations of extensive forest areas by species of the genus Acacia begin to be very frequent, but there are others, such as Hackea, that begin to occupy very significant areas and no longer only occupy space along the circulation ways.

However, as a consequence of climate change, forests can undergo more rapid changes, for example, unless there is a significant increase in precipitation, the severity of the fires may increase. For this reason, it is urgent to create a model of forest management that takes into account the phenomenon of climate change and all the variables associated with it, and never forget that forests are a re-course that can and must be monetized, in a perspective of sustainability for the future.

Printed in the United States
By Bookmasters